Lecture Notes in Mathematics 1622

Springer
Berlin
Heidelberg
New York
Barcelona
Budapest
Hong Kong
London
Milan
Paris
Santa Clara
Singapore
Tokyo

Emmanuel Dror Farjoun

Cellular Spaces, Null Spaces and Homotopy Localization

Springer

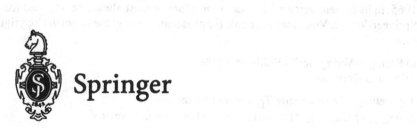

Author

Emmanuel Dror Farjoun
Mathematics Department
Hebrew University of Jerusalem
Jerusalem, Israel
E-Mail: farjoun@sunset.huji.ac.il

Cataloging-in-Publication Data applied for

Die Deutsche Bibliothek - CIP-Einheitsaufnahme

Farjoun, Emmanuel Dror:
Cellular spaces, null spaces and homotopy localization / Emmanuel Dror Farjoun. - Berlin ; Heidelberg ; New York ; Barcelona ; Budapest ; Hong Kong ; London ; Milan ; Paris ; Tokyo : Springer, 1995
(Lecture notes in mathematics ; 1622)
ISBN 3-540-60604-1
NE: GT

Mathematics Subject Classification (1991): 55

ISBN 3-540-60604-1 Springer-Verlag Berlin Heidelberg New York

Typesetting: Camera-ready $\TeX$ output by the author
SPIN: 11374442 46/3111-5431 – Printed on acid-free paper.

Contents

Introduction

In these notes we describe in some detail a certain framework for doing homotopy theory. This approach emerged in the early 1990's but has roots in earlier work of Bousfield about localization and in the big advances made by Mahowald, Ravenel, Devinatz, Hopkins and Smith towards deeper understanding of the role of periodicity in stable homotopy theory. It is natural to look for a similar unstable organization principle. This has not been found. Rather, certain tools have developed that have proved interesting. In addition, these tools are closely related to the above developments, as well as to central developments that occurred in unstable homotopy with the proof by Miller of the Sullivan conjecture and with the fruitful use of Miller's theorem by Lannes, Dwyer, Zabrodsky and many others.

During these developments the study of homotopy theory through function complexes has become common and productive. Computation of important function complexes has become possible, especially with classifying spaces as domains. It turns out that it is also very productive to formulate localization theory in terms of function complexes. In particular, the notion of a W-null space (essentially, a space X for which the pointed function complex $\mathrm{map}_*(W, X)$ is contractible) has become central in localization theory.

Thus function complexes play a central role in these notes. In fact one can view most of the material as developing techniques that allow better understanding of function complexes not via computing their homology or homotopy groups but directly as spaces. Therefore homotopy colimits become very useful, since it is convenient to have them as domains of function complexes. A typical situation is the decomposition of classifying spaces of compact Lie groups as homotopy colimits by Jackowsky, McClure and Oliver, which allowed a much deeper understanding of function complexes between these objects. In this framework we give an exposition of the work of Bousfield and Thompson about unstable localization and relate it to a better understanding of homological localization.

In relation to homotopy colimits a new tool that comes into play is that of cellular spaces. We show that these structures are closely related to localization, more specifically to colocalizations—homotopy fibres of the localization map. These structures are treated here as being of interest in their own right. They allow one to write, in some interesting instances, classical constructions as pointed homotopy colimits. For example, we examine the symmetric product SP^∞ in this light. This again allows one to better understand function complexes on these spaces which are decomposed as homotopy colimits.

Spaces, function complexes: The present notes can be read either in the category of topological spaces having the homotopy type of CW-complexes, or in the category of simplicial sets. We refer to both as 'spaces' and both categories are denoted by $\mathcal{S}$, or when we talk about pointed spaces as $\mathcal{S}_*$. Although it is perfectly possible to carry out almost the whole theory within the category $\mathcal{S}_*$ of (well-pointed) spaces we do not follow this path, since it is not always the easiest one (see 1.F.7). Rather we mix the discussion of the two categories, pointed and unpointed, trying to avoid the confusion that this might create. The category of simplicial sets is denoted by $\mathcal{SS}$ and that of topological spaces by Top. Often we use the notions of cofibrant and fibrant spaces. In Top cofibrant means (well-pointed) CW-complex while any space in $\mathcal{SS}$ is cofibrant. On the other hand, every topological space is fibrant while fibrant in $\mathcal{SS}$ means a simplicial set that satisfies the Kan extension condition [Q-1], [May-1]. Whenever some construction in Top, especially those involving mapping spaces, yields a non-CW space we can and do pull them back to the class of CW-spaces via the canonical CW-approximation (compare e.g. (1.B) or (1.F)). By a finite space we mean finite CW-complex or a simplicial set with a finite number of non-degenerate simplices.

Since we make extensive use of function complexes, care must be taken that simplicial sets that serve as ranges in function complexes are fibrant, satisfying the Kan extension condition [May-1], while spaces that serve as domains are always assumed to be cofibrant. Otherwise the homotopy type of a function complex is not invariant under weak equivalence and has in general no homotopy meaning. When we write $\mathrm{map}_*(X,Y)$ or $\mathrm{map}(X,Y)$ in the topological category we most often use only the underlying weak homotopy type of the space of continuous maps (pointed or unpointed), so there is no need to turn it into an internal function complex having the homotopy type of a CW-complex. For typographical reasons the notation Y^X is often used to denote the function complex of maps from X to Y. We denote by $\simeq$ a weak homotopy equivalence. Certain constructions though are easier to handle in the category of simplicial sets where $\mathrm{map}(X,Y)$ denotes the usual simplicial function complex [M-1]. It is often possible to carry over the necessary construction naturally into topological spaces using the pair of adjoint functors, the realization and singular functors. This is demonstrated in some detail in section 1.F.

A note about chapters and sections: References within the nine chapters are by sections, such as (B.3.5). When referring to other results or sections outside the current chapter, the number of the chapter precedes that of the section or result, e.g. (1.F.6.1) is a result or a figure from Chapter 1, section F.

Some details about the contents: In Chapter 1 the basic notions of f-local space and f-localization with respect to an arbitrary map denoted $\mathbf{L}_f$, are introduced. A special case, when the map f is null homotopic, has particularly pleasant properties and is called nullification, denoted by $\mathbf{P}_A$, when the map is $A \to *$. This last functor allows one to introduce an interesting partial order on spaces that is analyzed later on: one says that X 'supports Y' or 'kills Y', denoted by: $X < Y$, if $\mathbf{P}_X Y \simeq *$. This is really the same as the implication: For any space T, $\text{map}_*(X,T) \simeq * \Rightarrow \text{map}_*(Y,T) \simeq *$ (note, however, the different convention-notation followed in [B-4] where the sense of $<$ is reversed.)

We give a list of elementary properties of localization that forms the beginning of a sort of localization calculus, which will allow one to control the behavior of $\mathbf{L}_f$ under standard homotopy operations such as suspensions, loops, and homotopy colimits. These functors are universal in two senses: they are both terminal and initial up to homotopy in certain classes of maps. Still we do not know of any inverse limit constructions that present them as initial objects analogous, say, to the Bousfield–Kan construction of their localizations as an inverse limit.

We also begin to note some crucial properties that distinguish the nullification from the general localization. In particular the following seems to be a basic distinction:

When $\mathbf{P}_A$ is applied to the homotopy fibre of the coaugmentation map $X \to \mathbf{P}_A X$ one always gets a point up to homotopy: that is, there is a universal equivalence $\mathbf{P}_A(\text{Fib}(X \to \mathbf{P}_A X)) \simeq *$. The analogous formula for $\mathbf{L}_f$ is weaker.

Chapter 2 can be seen as an attempt to discuss more carefully the homotopy fibre of the nullification map. We now know that this homotopy fibre when considered as a functor on the pointed category of spaces is an idempotent augmented functor denoted by $\overline{\mathbf{P}}_A$. It is a sort of colocalization. Since $\mathbf{P}_A X$ is really X stripped of all its 'A-information' the homotopy fibre $\overline{\mathbf{P}}_A X$ still contains all this information, and in fact $\text{map}_*(A, \overline{\mathbf{P}}_A X)$ is equivalent to $\text{map}_*(A, X)$. But in general $\overline{\mathbf{P}}_A X$ is not the universal space with this property.

There is another canonical space denoted by $\mathbf{CW}_A X$, which is the universal space having the same function complex from A as X. Furthermore, this space is built out of copies of A and approximates X much in the same way that a classical CW-approximation (which is 'composed of cones on spheres' and extracts the 'spherical information' from X expressed in the usual form of the homotopy groups) gives a 'spherical approximation' to X. Thus we consider here a second partial order, denoted by $\ll$, which, as it turns out, is closely related to $<$ defined above: namely, $X \ll Y$ if and only if the pointed space Y can be built from the pointed space X by repeatedly applying, say, wedges and homotopy pushouts, possibly infinitely many times. We say that Y is X-*cellular* in that case and we begin to consider the above cellularization functor and this partial order in this chapter. For example, one shows that a finite product of X-cellular spaces is X-cellular (2.D.16). Notice that if $X \ll Y$, and X is acyclic with respect to any homology theory, then so is Y (2.D.2.4). Also we shall see that $X \ll Y$ always implies $X < Y$.

In particular we begin to develop criteria to decide when a given space X is A-cellular with respect to another space A, i.e. under what conditions the equivalence $X \simeq \mathbf{CW}_A X$, or equivalently $A \ll X$, holds. Such criteria are an important concern in these notes. This is handy when, for example, one wants to know under which conditions a K-acyclic space can be constructed by pushouts and telescopes from elementary K-acyclic spaces such as the cofibre of the Adams map, and is therefore the direct limit of its K-acyclic finite subspaces.

We then see that we have obtained two seemingly closely related functors. One would like to show, but it is not yet known how to proceed in all cases, that these two idempotent functors, the localization $\mathbf{L}_f$ and cellularization $\mathbf{CW}_A$, are in fact two facets of a symmetric construction that factors an arbitrary map $X \to Y$ into a 'cofibration' followed by a 'trivial fibration': One can change the usual notions of weak equivalences in the 'standard model category' of spaces by, say, adding a single map f to the class of weak equivalences, but this will change also the notion of fibre maps along which one should be able to lift weak equivalences that are cofibrations. The localization would then be just factorization of the map $X \to *$ while cellularizations are factorizations of the map $* \to Y$. These observations put the above functors in a reasonable theoretical light. Hirschhorn is developing these directions carefully in a general framework [HH]. We then continue to show that certain standard constructions lead to cellular relations: For example, the cellularity of the third term in a fibration sequence can be predicted when one knows the other two.

In Chapter 3 we turn to deeper technical properties of these idempotent functors: the rule of commutation with the loop space functor and in general with taking the homotopy fibre of a map. It turns out that one has general formulas $\mathbf{L}_f\Omega \simeq \Omega\mathbf{L}_{\Sigma f}$ and $\mathbf{CW}_A\Omega \simeq \Omega\mathbf{CW}_{\Sigma A}$. These are a fundamental part of the calculus of localization and are used to show, for example, that fibrations over a W-null base space are preserved by nullification with respect to the suspension of W, but also much more general theorems concerning preservation of fibrations.

These formulas also allow us to compute directly, and formally, certain cellular relations, such as the fact that $\Omega\Sigma X$, the James construction on X, is always X-cellular. In fact one can show that $\Sigma X < Y \Rightarrow X \ll Y$.

Chapter 4 serves two purposes. First, we present a more careful analysis of pointed homotopy colimits and their relations to the usual strict colimits (direct limits) of diagrams of spaces. This allows us to show, for example, that $SP^\infty X$, the Dold–Thom symmetric product on X, which initially, like the James construction, is defined as a strict colimit via a point-wise construction, can in fact be built by a pointed homotopy colimit starting from the initial space X alone. Since by the Dold–Thom theorem the infinite symmetric product is a GEM, i.e. a product of

Eilenberg–Mac Lane spaces, and is, in fact, the universal GEM associated with X, its expression as a homotopy colimit allows one to understand better the operation of localization and cellularization on generalized Eilenberg–Mac Lane spaces. This paves the way to the second purpose: a cellular version of a 'key lemma' of Bousfield. This version, following an approach taken by Dwyer [Dw-2], describes the cofibre of the map from the Borel construction on X to the corresponding strict quotient space. The key lemma is related to a cellular estimate of the cofibre as being $\Sigma^2 X$-cellular. This means, roughly speaking, that in order to build $\Sigma SP^\infty X$ from ΣX, exactly one copy of ΣX is needed, and then only higher suspensions of X.

In Chapter 5 we show that if in a fibration sequence $F \longrightarrow E \longrightarrow B$, $\mathbf{P}_{\Sigma A}$ kills both the base space and the total space, i.e. $\mathbf{P}_{\Sigma A} B \simeq *$ and $\mathbf{P}_{\Sigma A} E \simeq *$, it may not kill the fibre F, but it always turns it into an 'homotopy abelian object': we show that $\mathbf{P}_{\Sigma A} F$ is naturally an infinite loop space that is equivalent as such to a product of Eilenberg–Mac Lane spaces with their usual abelian infinite loop space structure. This is done using the results of Chapter 4, and notably Bousfield's key lemma and the 'infinite loop space machine' of Segal, as well as substantial parts of [DF-S], extending their results to cofibrations and the cellular approximation functor.

This approach leads to a general theorem about the preservation of fibration by the nullification 'up to an abelian error term'. We then use the relation between localization with respect to a map and nullification with respect to its cofibre to deduce a general theorem about the localization of arbitrary fibration with respect to any double suspension. It is perhaps worth mentioning here that the fact that the classical Sullivan type of localization preserves fibrations over, say, a 1-connected space is a special case of these preservation-of-fibrations theorems: from the present point of view, the Sullivan localization of a simply-connected space with respect to a prime p is just the Anderson localization [An], which is in turn simply nullification with respect to $\Sigma M^2(p)$, the suspension of the two-dimensional Moore space. Similar reasoning is then applied in examining the effect of applying $\mathbf{CW}_A$ to a fibre sequence, with results that are weaker, but similar to the above.

We then apply this theory to show two remarkable examples: first, we describe a theorem of Neisendorfer which states that any finite 2-connected complex can be recovered, up to p-completion, from its n-connected cover, for any $n \geq 0$. This is much stronger than saying that these spaces must have non-trivial homotopy in infinite dimensions: it shows that somehow this 'infinite tail' has all the information needed to reconstruct the 'lower dimensional information'. Secondly, we show that this Serre-type result about homotopy groups in high dimensions generalizes to infinite spaces, too: their A-homotopy groups must be non-trivial in infinitely many dimensions, as long as these spaces are built by a finite number of A-cells from $\Sigma^i A$, where $i \geq 1$.

In Chapter 6 we turn our attention to homological localization with respect to generalized homology theories, such as Morava K-theory. We essentially reproduce the relevant material from [DF-S], showing that short of a small error term these localizations preserve twice looped fibrations. It is reasonable to expect that this is also the case for single-loop-space fibrations.

Note, however, that in order to present homological localization in the form $\mathbf{L}_f$, for some map f, one needs to take a 'monster map': in general, one must take the union of all E_*-homology isomorphisms between spaces whose cardinality is not bigger than the coefficients of E_*.

While theoretically this can be done, it is certainly desirable to replace this 'monster map' with a smaller object. To do that so, one considers a classification of possible nullification functors, under some restrictions in Chapter 7. This means roughly the classification of all possible nullification functors with respect to finite p-torsion suspension spaces. The above mentioned nullity classes of spaces define a very rough equivalence on spaces, namely, W and V are null-equivalent or of the same nullity class if, for any pointed space T, one has the double implication: $\mathrm{map}_*(V,T) \simeq *$ if and only if $\mathrm{map}_*(W,T) \simeq *$. This really means that the functors $\mathbf{P}_W$ and $\mathbf{P}_V$ are naturally equivalent. The classification of these classes, starting with similar but much easier stable classes, is undertaken in Chapter 7. It turns out, making heavy use of Bousfield's theory of fibrations above, that the stable and unstable classifications are not very different from each other, and the main invariant needed here is the stable one, namely, the 'Hopkins–Smith type' related to Morava K-theories. This possibly is not all that surprising since, if one localizes with respect to a nilpotent self-map $\Sigma W \to W$, one obtains the same results as nullification with respect to W and, by [D-H-S], there is essentially only one self-map on the above complexes that is not nilpotent. Using the classification of nullity classes one can also classify the closely related cellular classes of the above suspension spaces.

The classification of nullity classes and possible nullification functors can be used to analyse higher periodicity. This is done in Chapter 8 with respect to v_1. We follow the work of Bousfield and Thompson in regard to K-theory localization. But we use the cellular analysis to express the nullification and K-localization functors in more elementary terms, as a telescope of the Adams map. Thus in this case, modulo some technicalities, the above monster map used to express K-localization can be replaced by a single map between two Moore spaces. This is a sort of ideal situation that could hold, in general, if a kind of 'unstable telescope conjecture' were true.

The basic result is that one can express the function complex $\mathrm{map}_*(M, \mathbf{P}_{V(1)}E)$ as a mapping telescope of function complexes, which inverts v_1

in the most elementary way: namely, as

$$\mathrm{Tel}\left(E^M \to E^{\Sigma^q M} \to E^{\Sigma^{2q} M} \to \cdots\right)$$

where M denotes the appropriate Moore space. This latter telescope is of course neither idempotent nor coaugmented as a functor, so it cannot replace the localization in general, but it still captures in a direct way the v_1-periodic homotopy. In particular its homotopy groups depend only on the action of the v_1 operator on $\pi_* E$ as a graded group.

As a corollary one can explain to what extent higher loop spaces on K-acyclic spaces are still acyclic. In addition one can show that K-acyclic, p-torsion spaces whose loop spaces are also K-acyclic can be built via cofibration sequences from the cofibre of the Adams map v_1.

In the final chapter, Chapter 9, we develop several tools that allow us to detect and prove interesting cellular inequalities. The basic idea here is the passage from *pointed* homotopy colimits over arbitrary indexing categories to *unpointed* homotopy colimits over categories with a contractible classifying space. Explaining ideas from [Ch-2] and [DF-5], the program here is to show that often the well-known theorems on connectivity of homotopy constructions such as homotopy fibre can be strengthened to a theorem asserting the cellularity of these constructions. On occasion this gives new connectivity results too. A typical result in this direction is that the fibre of the cofibration quotient map $X \to X/A$ is always A-cellular. Furthermore the homotopy fibre of a map to connected space can be built by pointed homotopy colimits from the collection of the actual inverses of points (say barycentres) in the base space.

We conclude with some applications of this technique: In particular one can show that while there is no easy relationship between the homotopy limit and colimits of diagrams, some inequalities can be proven in general between these constructions. These form a sort of generalization of the elementary inequality $\Sigma\Omega X \ll \Omega\Sigma X$.

Acknowledgments: Several crucial ideas that I have tried to explain here were developed by A.K. Bousfield in his paper [B-4] about unstable periodicity. In particular he proved there the key lemma that opens the way to many of the more interesting results about localizations. This work was initiated while I was visiting the University of Geneva and the E.T.H. research institute and I am grateful for their hospitality. A good part of the present work was done while I was visiting Purdue University. I am very grateful for their hospitality and the close cooperation in particular with Jeff Smith with whom [DF-S], which is also a part of the present notes, was written. I am very grateful to W. Dwyer for his help in deciphering the relations of the key lemma to homotopy coends and his proof of (3.C.16). His support in the project of understanding cellular spaces was very useful. It was partly done in the framework of a research grant from the Binational Science Foundation. No less useful was the close cooperation on the initial chapters with Hirschhorn who is now developing the theoretical foundations and model theoretical development in a separate monograph. Dwyer's student W. Chacholsky has taken an interest in these problems and greatly helped in developing ideas about cellularity of homotopy fibres in Chapter 9. The referees' comments were very valuable and helpful. I am also very grateful to TEX specialist Simcha Kojman who was very patient with my changes; she and copy editor Paul Greenberg greatly facilitated the final production.

1. COAUGMENTED HOMOTOPY IDEMPOTENT LOCALIZATION FUNCTORS

Introduction

In this chapter basic notions that will be used throughout the present notes are defined, in particular that of f-local spaces and f-equivalences between spaces. A list of elementary properties of the basic notions and of localization is given in section 1.A.8 below. These properties mostly follow easily and directly from the definitions and are used in many arguments. A construction of the f-localization functor is given and its property of continuity is discussed. This is useful but not essential in constructing a fibrewise version of localization. Continuity also renders certain induced maps such as $\text{aut}(X) \longrightarrow \text{aut}(\mathbf{L}_f X)$ easily understandable, where aut denotes the spaces of self-homotopy equivalences.

We pay some attention to the localization of homotopy colimits. The interested reader can find a brief discussion of these colimits in Appendix HL below. The fact that localization behaves relatively well under homotopy colimits–including wedge sum, for example–is very helpful later on. We then show that well-known localization functors including e.g. the Quillen plus construction are special cases of this general homotopy localization. Then there is a discussion of fibrewise localization and several approaches are discussed. This discussion is carried out in a bit more general framework of applying homotopy functors fibrewise. We show how to do that using homotopy colimits under mild assumptions on the functor. As a first application of these fibrewise localizations one deduces two very useful properties of the homotopy fibre of the localization map: first we show that if the localization kills the fibre then it preserves the fibration. Then we show that the localization with respect to a null map always kills the homotopy fibre of the localization map.

A. Local spaces, null spaces, localization functors, elementary facts

We consider here the notion of f-local space where $f : A \to B$ is an arbitrary cofibration map between cofibrant spaces (i.e. CW-complexes if we work in Top). Bousfield in [B-2] has already shown how to associate an f-local space $\mathbf{L}_f X$ with any space X together with a coaugmentation map $X \to \mathbf{L}_f X$. It turns out that in spite of its generality, this localization functor has many useful properties that combine to form a 'calculus of localization'. We examine in this chapter some of these most basic properties. In case the cofibration f is a null homotopic map, the functor $\mathbf{L}_f$ has stronger and cleaner properties and is called *nullification* (with respect to $A \vee B$, see below).

In practice, all the known coaugmented homotopy functors **F** which are also idempotent (i.e. roughly **FF** is equivalent to **F**) have the form $\mathbf{L}_f$ for a suitable f, so the present framework and result may well apply to any idempotent functor. Notice, however, that Bousfield–Kan's R_∞ is not in general idempotent.

In this chapter we also consider somewhat more delicate properties of localization, in particular its value on homotopy colimits and on the homotopy fibre of the localization (coaugmentation) map.

A.1 DEFINITION : *(f-local, W-null): We say that Y is f-local (where f is a map $f : A \to B$ between cofibrant spaces) if Y is fibrant and the map f induces a weak homotopy equivalence on function complexes,*

$$\mathrm{map}(f, Y) : \mathrm{map}(B, Y) \xrightarrow{\simeq} \mathrm{map}(A, Y).$$

In case the map is simply $w : * \to W$ one refers to a w-local space Y as W-null; this means that the natural map $Y \xrightarrow{\simeq} \mathrm{map}(W, Y)$ is an equivalence. Equivalently one defines these concepts in the pointed category of spaces (where now all spaces are assumed to be well-pointed): A fibrant space is local if the corresponding map of function complexes of pointed maps is a weak equivalence

$$\mathrm{map}_*(f, Y) : \mathrm{map}_*(B, Y) \xrightarrow{\simeq} \mathrm{map}_*(A, Y).$$

Remark: The fibration $\mathrm{map}_*(V, X) \longrightarrow \mathrm{map}(V, X) \longrightarrow X$ for any cofibrant V over any connected and fibrant X shows that for a connected and fibrant space X the map induced by f, namely $\mathrm{map}(f, Y)$, is an equivalence iff the map $\mathrm{map}_*(f, Y)$ is an equivalence with respect to any choice of ('well-pointed') base points.

A.1.1 EXAMPLES: We give examples in sections E and 2.D below. Here we note several quick illustrations: If the map f is the map of the n-sphere to a point $* \longrightarrow S^n$, then an f-local connected space is an S^n-null connected space, i.e. it is a space X whose n-th loop space $\Omega^n X$ is contractible. Thus such a space has no homotopy groups above dimension $n-1$ and is otherwise arbitrary. Thus it is just an arbitrary Postnikov $(n-1)$-stage. Dually it is easy to see that a space X is n-connected if and only if any Eilenberg–Mac Lane space $K(G, i)$ for $0 \le i \le n$ is X-null. For a more difficult example, if the map g is the degree p map from the n-sphere to itself, then a *connected, pointed* space X is g-local if the map on its n-th loop space raising every loop to its p-th power is a weak homotopy equivalence of the underlying spaces, disregarding the loop structure. For $n > 1$ this means that all the homotopy groups above dimension $(n-1)$ are uniquely p divisible, see E.3 below.

A.1.2 REMARK: One might ask why we do not define a 'homotopy f-local' space to be a space W for which the induced map on homotopy classes of maps (rather than on the full function complexes), namely the map $[f, W] : [B, W] \to [A, W]$ is an isomorphism of sets. This is a perfectly good definition, but it turns out that it too leads directly to the definition given above. The reason is that given a notion of 'f-local' space we are mostly interested in functors that turn an arbitrary space into an 'f-local' one. Now the following fact shows that as far as functorial constructions are concerned the definition using homotopy classes leads to one in which the full function complexes are used.

A.1.3 FACT: For any continuous (or simplicial, see C.8 below), idempotent, coaugmented functor F : {Spaces} $\to$ {Spaces}, if, for all X, the induced map on homotopy classes $[f, FX]$ is an isomorphism of sets, then FX is automatically f-local: $\text{map}(f, FX) : \text{map}(B, FX) \xrightarrow{\sim} \text{map}(A, FX)$ is an equivalence.

This is Corollary (1.3) in [DF-4] which was written in view of this and similar questions.

Another way of viewing Fact A.1.3 is to notice that it implies that it is impossible to canonically associate a universal 'homotopy f-local' space with every space X.

This is best understood by an example (due to G. Mislin): Let $f : S^1 \to *$, then a 'homotopy f-local' space is just a simply connected space. We ask: Is there an initial object among all maps of a space, say of $\mathbb{R}P^2$, to 1-connected spaces? The answer is NO. To see why, notice that by unique factorization up to homotopy such a space U would need to have $H^2(U, \mathbb{Z}) \cong \mathbb{Z}/2\mathbb{Z}$ since the non-trivial map $\mathbb{R}P^2 \to \mathbb{C}P^\infty$ would also have two factors through U, uniquely up to homotopy. But U is 1-connected and its second cohomology cannot have torsion.

A.2 DEFINITION: *A functor* $\mathbf{F}$ *is called* coaugmented *if it comes with a natural transformation* Id $\to \mathbf{F}$, *i.e. for each* $X \in \mathcal{S}$ *a natural map* $j_X = j : X \to \mathbf{F}X$. *A coaugmented functor* $\mathbf{F}$ *is said to be idempotent if both natural maps:* $\mathbf{F}X \rightrightarrows \mathbf{F}\mathbf{F}X$, *namely both* $j_{\mathbf{F}X}$ *and* $\mathbf{F}(j_X)$, *are weak equivalences and are homotopic to each other. We say that the coaugmentation map* j_X *is* homotopy universal *with respect to maps* $X \to T$ *into* f*-local spaces* T *if any such map factors up to homotopy through* $X \to \mathbf{F}X$ *and the factorization is unique up to homotopy.*

The next few pages will present a construction of localization functor [B-2] [DF-2] [C-P-P]:

A.3 THEOREM: *For any map $f : A \to B$ in $\mathcal{S}$ (or $\mathcal{S}_*$, see remark A.7) there exists a functor $\mathbf{L}_f$, called the f-localization functor, which is coaugmented and homotopically idempotent. Any two such functors are naturally weakly equivalent to each other. The map $X \to \mathbf{L}_f X$ is a homotopically universal map to f-local spaces. Moreover, $\mathbf{L}_f$ can be chosen to be continuous or simplicial in the sense explained (1.C) below.*

Proof: The construction of $\mathbf{L}_f$ is carried out in section B below. The proofs of claims about $\mathbf{L}_f$ are in (B.5), (C.1), (C.2), and (C.12) below.

A.4 NULLIFICATION FUNCTORS $\mathbf{P}_W$, NULLITY CLASSES: A special role is played by localization with respect to maps of the form $W \to *$, or $* \to W$. In that case a pointed and connected space X is $(W \to *)$-local or, by (A.1), W-null if and only if $\mathrm{map}_*(W, X) \simeq *$ or $\mathrm{map}(W, X) \simeq X$. The localization with respect to these null maps deserves a special name due to its much better behavior and common occurrence. One denotes the localization $\mathbf{L}_{W \to *} = \mathbf{L}_{* \to W}$ by $\mathbf{P}_W$; we call $\mathbf{P}_W$ the W-nullification functor. Bousfield used the term W-periodization for $\mathbf{P}_W$. It plays a major role in his theory of unstable periodic homotopy, as we shall see below. This notation also emphasizes the affinity of general nullification functors to their early predecessor, the Postnikov section functor P_n that we saw above.

Of course, the condition $\mathrm{map}_*(W, X) \simeq *$ occurs often in homotopy theory especially since the proof by Miller of the Sullivan conjecture, that says in these terms that any finite-dimensional space is $K(\pi, 1)$-null for any locally finite group π. The concept of trivial function complex plays a major role in the present notes and we use it right away to define a useful partial order on pointed or unpointed spaces.

A.5 NULLITY CLASSES, (WEAK) PARTIAL ORDER $X < Y$: We say that X supports Y or that Y is X-supported and denote it by $X < Y$ if any X-null space is also Y-null. This is a transitive but not anti-reflexive relation. It is equivalent as we shall see to $\mathbf{P}_X Y \simeq *$ (A.8)(e.9) below. One says that X and Y have the same nullity (class) if $X < Y$ and also $Y < X$. Thus $S^n < S^{n+1}$ and $X \vee X$ has the same nullity as X. Notice that [B-4] uses the opposite convention in the notation of the partial order

A.6 EXAMPLE: $P_{S^{n+1}}$ is the n-th Postnikov section $P_n X$ which can be characterized by $\Omega^{n+1} P_n X \simeq *$. Compare A.1.1 above. An important result of Zabrodsky and Miller [M], a strong version of which is given in (2.D.13) below, says in this notation that for any topological group G one has $G < BG$. In fact we shall prove the sharper inequality: $\Sigma G < BG$ is always true. See (9.D.4) below.

A.7 Pointed and unpointed spaces: Notice that in Definition A.1 above we considered the function complex map(,) of an unpointed map between two unpointed spaces. We have noticed there that by the same token one can consider the notion of f-local pointed spaces in $\mathcal{S}_*$ where $f : A \to B$ is a map in $\mathcal{S}_*$. As we noticed, if X is a connected pointed space, then X is f-local in $\mathcal{S}_*$ if and only if it is f-local in $\mathcal{S}$—after forgetting the base points. One advantage of the unpointed version is that it allows the direct construction of fibrewise localization (F.1). We will see that we get a continuous or simplicial functor (see section C below) and thus it can be applied to each fibre. Since there is no continuous choice of base points across the fibres, we cannot directly use the pointed functor here. If one restricts attention to connected spaces then this is just a convenience because the value of localization functors pointed or unpointed are the same for connected spaces.

A.8 Elementary facts concerning f-local spaces and $\mathbf{L}_f$: The following are almost immediate consequences of the definitions, universality, idempotency and (A.3) above and are used implicitly in many arguments. We will give their proofs in section G below. As usual whenever discussing a function complex one assumes that the range is cofibrant and the target is fibrant.

- e.1 For any two maps f, g we have $\mathbf{L}_g \simeq \mathbf{L}_f$ if and only if any f-local space is g-local and vice versa.
- e.2 If T is f-local, then for all (well-pointed, cofibrant) X both $\mathrm{map}(X,T)$ and $\mathrm{map}_*(X,T)$ are f-local for any choice of base points. In particular, if T is f-local so is $\Omega^n T$ for all $n \geq 0$.
- e.3 More general than (e.2) is the observation that any homotopy limit of f-local spaces is f-local. In particular, the homotopy fibre of a map between two connected f-local spaces is f-local; the product of any family of f-local spaces is f-local.
- e.4 It follows easily from the exponential law and universality that the natural map $\mathbf{L}_f(X \times Y) \to \mathbf{L}_f X \times \mathbf{L}_f Y$ has a homotopy inverse and thus is a homotopy equivalence.
- e.5 Let f, g be pointed maps. If $\mathbf{L}_f \simeq \mathbf{L}_g$ and $W \in \mathcal{S}_*$ any space, then $\mathbf{L}_{f \wedge W} \simeq \mathbf{L}_{g \wedge W}$. This is because the assumption implies that any $T \in \mathcal{S}_*$ is $(f \wedge W)$-local iff it is $(g \wedge W)$-local by use of adjunction.
- e.6 If $F \to E \to X$ is a fibration over a connected X, and both F and X are W-null (A.4), then so is E. This is because, in general, the induced sequence on function complexes $F^W \to E^W \to X^W$ is also a fibration sequence, where F^W is the fibre over the null component of X^W.
- e.7 A connected space $T \in \mathcal{S}_*$ is local with respect to $\Sigma f : \Sigma A \to \Sigma B$ if and only if (ΩT) is f-local. Thus, if in a fibration sequence, as in (e.6), F is f-local and B is Σf-local, then E is also Σf-local. More generally, let X be local with respect

to the n-th suspension of f and let W be any $n+1$ connected space. Then the pointed function complex $\text{map}_*(W, X)$ is local with respect to f itself. This follows from the adjointness of Σ and Ω: $\text{map}_*(\Sigma f, T) = \text{map}_*(f, \Omega T)$.

e.8 If T is f-local, then it is also $\Sigma^k f$-local with respect to any suspension of f, i.e. for all $k \geq 0$.

e.9 $\mathbf{L}_f X \simeq *$ if and only if for any f-local space P one has $\text{map}_*(X, P) \simeq *$.

e.10 If $\mathbf{L}_f X \simeq *$ then also $\mathbf{L}_{\Sigma f}(\Sigma X) = \mathbf{L}_{\Sigma f}(S^1 \wedge X) \simeq *$. More generally if, in addition, W is any n-connected space, then $\mathbf{L}_{\Sigma^{n+1} f}(W \wedge X) \simeq *$.

e.11 If $f : A \to B$ is a map of n-connected spaces for some $n \geq 0$ and X is also n connected, then so is $\mathbf{L}_f X$.

B. Construction of $\mathbf{L}_f$

The following construction of the homotopy localization $\mathbf{L}_f$ can be read simplicially or topologically. In carrying through the construction in the topological category the present construction does not remain inside CW-complexes. But one can always push it into the CW category in a functorial way as follows, using a natural CW approximation $CW(Y) \to Y$:

Consider the following natural square.

$$\begin{array}{ccc} CW(X) & \longrightarrow & CW(\mathbf{L}_f X) \\ \downarrow & & \downarrow \\ X & \longrightarrow & \mathbf{L}_f X \end{array}$$

We can assume that the coaugmentation map is a cofibration (otherwise we can functorially turn it into one) and so is the induced map on the CW approximations. If $\mathbf{L}_f^{cw} X$ is the pushout then there is a canonical factorization $X \to \mathbf{L}_f^{cw} X \to \mathbf{L}_f X$, in which the second map is a weak equivalence. Also, if X is a CW complex then so is the pushout. In this way we have defined a localization map canonically inside the category of CW complexes.

We construct $\mathbf{L}_f X$ for any X as the colimit of a transfinite tower of cofibrations:

$$X = \mathbf{L}^0 X \hookrightarrow \mathbf{L}^1 X \hookrightarrow \cdots \mathbf{L}^\beta X \hookrightarrow \cdots \quad (\beta < \lambda)$$

for a certain ordinal $\lambda = \lambda(A \amalg B)$. We prove two crucial properties of $\mathbf{L}_f X$: First, the easier part if X is f-local then the map $X \to \mathbf{L}_f X$ is a (weak) homotopy equivalence, see (B.2). Second, for all X we have $\mathbf{L}_f X$ is f-local, see (B.5). These two will be used to show that $\mathbf{L}_f$ satisfies the claims of (A.3). (See Section C below.)

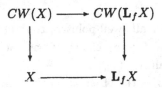

REMARK: The present section can be viewed as a version of section (5) of [B-3] which treats homological localizations.

The construction is in essence a 'continuous' adaptation of the 'small object argument' of Quillen [Q-1].

B.1 SMALL OBJECT ARGUMENT, RECOGNITION OF LOCAL SPACES

It might help to put the construction in perspective. Working as we do in the unpointed category one starts with an <u>unpointed</u> criterion for establishing that a map is a weak equivalence. Here we use the basic relative cell or simplex: $\dot{\Delta}[n] \hookrightarrow \Delta[n]$, namely the inclusion of the boundary of the n-simplex into the n-simplex $\Delta[n]$

FACT: a map $X \longrightarrow Y$ is a weak equivalence fibre map if and only if for any pair of maps $\bar{\delta}$, δ that renders the following square diagram (strictly) commutative, i.e. with $\delta \circ i = g \circ \bar{\delta}$, there is a diagonal lifting ℓ that renders the diagram commutative.

$$\begin{array}{ccc} \dot{\Delta}[n] & \xrightarrow{\bar{\delta}} & X \\ {\scriptstyle i}\downarrow & \nearrow{\scriptstyle \ell} & \downarrow{\scriptstyle g} \\ \Delta[n] & \xrightarrow{\delta} & Y \end{array}$$

Thus, for example, for $n = 0$ one has $\dot{\Delta}[0] = \emptyset$, this means that the map g is surjective on the vertices or on points, this together with the condition for $n = 1$ implies that g induces isomorphism on path components, etc.

Given a cofibration $f : A \hookrightarrow B$ we would like to turn a given space X into an f-local space $\mathbf{L}X$. Such a space $\mathbf{L}X$ would be fibrant and satisfy $f^* \equiv \text{map}(f, \mathbf{L}X) : (\mathbf{L}X)^B \xrightarrow{\simeq} (\mathbf{L}X)^A$ is a weak equivalence. Since we assume that f is a cofibration we get the induced map f^* is a fibre map [Sp], (assuming $\mathbf{L}X$ is fibrant.) So we need that the induced map $\text{map}(f, \mathbf{L}X)$ would be a fibre map that is a weak equivalence. Thus we use the criterion above to recognize whether the map $(\mathbf{L}X)^B \longrightarrow (\mathbf{L}X)^A$ is an equivalence.

Applying the criterion above, one gets, by a simple application of exponential law of function complex, that a given space Y is f-local if for any pair of maps $\bar{\delta}_1$,

δ_1 in the following diagram:

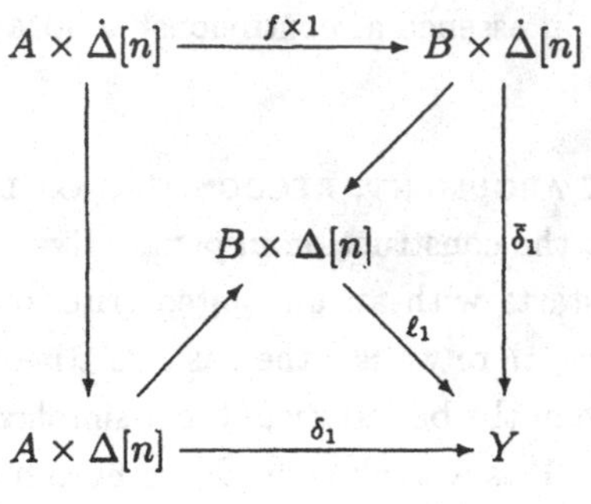

There is an extension $\ell_1 : B \times \Delta[n] \to Y$ that renders the whole diagram strictly commutative.

Notice that in the above diagram all arrows not involving Y are given a-priori by the inclusion $\dot{\Delta}[n] \hookrightarrow \Delta[n]$ and the cofibration f.

Rearranging the spaces and taking union over the upper left triangle we may re-write the condition on Y to be: Y is f-local if for any map g in the following diagram:

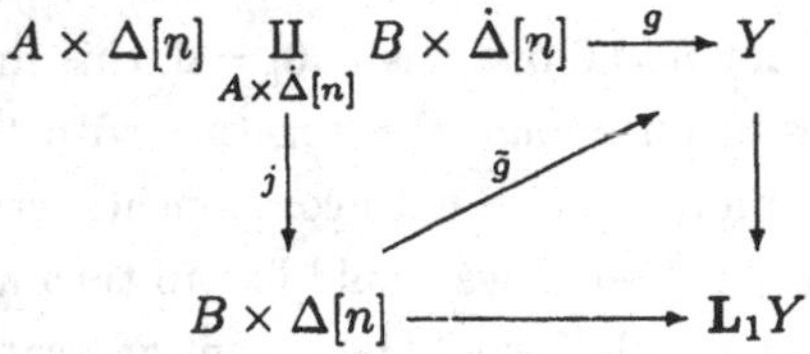

there exists an extension $\tilde{g} : B \times \Delta[n] \to Y$ rendering the upper triangle commutative: $g = \tilde{g} \cdot j$. The map j is given by the coherent inclusions. Now for a general Y, $\tilde{g}$ does not exists, therefore we change Y into $\mathbf{L}_1 Y$ by taking the (homotopy) pushout of j and g to be $\mathbf{L}_1 Y$, which is also the homotopy pushout since j is a cofibration. Replacing g by the disjoint union over all possible maps g and then taking the pushout we get the map $Y \to \mathbf{L}_1 Y$ which is the induction step of constructing $\mathbf{L}Y$, a local space associate to Y. One needs to continue this step since $\mathbf{L}_1 Y$ itself will not be local in general. But repeating the step by taking $\mathbf{L}_1(\mathbf{L}_1 Y)$, $\mathbf{L}_1(\mathbf{L}_1(\mathbf{L}_1 Y)))$ etc. transfinitely many times for a cardinal that is large than the 'size' of $(A \amalg B) \times \Delta[n]$ the latter become a 'small object' compare to the size of the tower $(\mathbf{L}_\beta Y)_\beta$ and the transfinite direct limit leads to an f-local $\mathbf{L}_f Y$ space as needed. This concludes our rough sketch of the transfinite use of the small object argument. The construction below is essentially a continuous version of this basic small object argument.

Let us now carry on with the actual construction of the functor $\mathbf{L}_f$. The advantage of the present approach is that it gives a built-in inverse on the function space level to the map induced by f, as well as a continuous functor. A version of this approach was first discussed by V. Halperin [H]. Given a map $f : A \to B$ we construct, by transfinite induction, for all ordinals $\beta \geq 0$, spaces $\mathbf{L}^{\beta}X$ together with natural transformations j that are all cofibrations (we omit the fixed function f from our notation here):

$$X = \mathbf{L}^0X \xrightarrow{j} \mathbf{L}^1X \xrightarrow{j} \mathbf{L}^2X \to \cdots \to \mathbf{L}^{\beta}X \to \mathbf{L}^{\beta+1}X \to \cdots.$$

The functor $\mathbf{L}^{\beta}X$ comes equipped with a natural map of function complexes: $s_{\beta} : (\mathbf{L}^{\beta}X)^A \to (\mathbf{L}^{\beta+1}X)^B$ that will serve in the limit as a (weak) homotopy inverse to the obvious induced map of function complexes: $(\mathbf{L}^{\lambda}X)^B \to (\mathbf{L}^{\lambda}X)^A$ induced by composition from $f : A \to B$. Here as elsewhere we denote, for brevity, by X^Y the function complex map(Y, X). The construction of $\mathbf{L}^{\beta}X$ is designed to render the following diagram commutative up to homotopy (see (B.4) below) where the maps s_{β} are defined following diagram (B.1.2) below. This will give, at the limit, the homotopy inverse to f^*.

(B.1.1)
$$\begin{array}{ccccc}
(\mathbf{L}^{\beta}X)^B & \xrightarrow{j^B} & (\mathbf{L}^{\beta+1}X)^B & \xrightarrow{j^B} & (\mathbf{L}^{\beta+2}X)^B \cdots \\
\downarrow f^* & \nearrow s_{\beta+1} & \downarrow f^* & \nearrow s_{\beta+2} & \\
(\mathbf{L}^{\beta}X)^A & \xrightarrow{j^A} & (\mathbf{L}^{\beta+1}X)^A \cdots & &
\end{array}$$

The inductive construction of $\mathbf{L}^{\beta}_f$

We define $\mathbf{L}^0X = X$ and, if β is a limit ordinal, we take $\mathbf{L}^{\beta}X$ to be the colimits of $\mathbf{L}^{\alpha}X$ for $\alpha < \beta$. Assume that $\mathbf{L}^{\beta}X$ is given. Define $\mathbf{L}^{\beta+1}X$ as $\mathbf{L}^1\mathbf{L}^{\beta}X$ where $\mathbf{L}^1X$ is the homotopy pushout in the following square. If one works with simplicial sets then, then, in order to get homotopy invariance, we take $\mathbf{L}^1X$ to be the Kan complex extension Ext^{∞} of the following homotopy pushout diagram applied to $\mathrm{Ext}^{\infty}X$ rather than to X itself:

(B.1.2)
$$\begin{array}{ccc}
A \times X^A \coprod_{A \times X^B} B \times X^B & \xrightarrow{f \amalg f^*} & B \times X^A \\
\downarrow ev & & \downarrow s_1' \\
X & \xrightarrow{j} & \mathbf{L}^1X
\end{array}$$

Here the union at the upper left is the obvious identification space which is both the colimit of $A \times X^A \leftarrow A \times X^B \rightarrow B \times X^B$ and its homotopy pushout, since $f : A \hookrightarrow B$ is a cofibration. The top map is the obvious map induced by f while ev is the evaluation. The map s_β mentioned above is induced by s_1' at each stage, where the map $s_1 : X^A \dashrightarrow (\mathbf{L}^1 X)^B$ in (B.1.1) for $\beta = 1$ is the adjoint to the map s_1' in the homotopic pushout square. The homotopy commutativity of diagram (B.1.1) above is immediate from the adjunction in this homotopy commutative pushout square see (B.4) below. We define $\mathbf{L}_f X$ to be $\mathbf{L}^\lambda X = \mathbf{L}_f^\lambda X$ for the ordinal $\lambda = \lambda(A \amalg B)$ chosen in (B.3) below.

The following two basic facts are immediate consequences of the construction: $\mathbf{L}_f X$ is a homotopy functor, and it does not change up to homotopy f-local spaces:

B.2 PROPOSITION: *(i) If the map φ: $X \longrightarrow Y$ is a weak equivalence then so is* $\mathbf{L}_f(\varphi)$

(ii) If X is an f-local space, then $X \rightarrow \mathbf{L}_f^1 X$ is a weak homotopy equivalence and thus so is $X \rightarrow \mathbf{L}^\beta X$ for all β.

Hence for such X we have a weak equivalence $X \rightarrow \mathbf{L}_f X$. In particular $\mathbf{L}_f(*) \simeq *$.

Proof: Each step of the construction is a homotopy functor i.e. preserves weak equivalences thus the final steep too. It is clear from elementary properties of homotopy pushout that j in (B.1.2) is a homotopy equivalence if $f \amalg f^*$ in the pushout square is such. But since the space at the upper left corner is a homotopy pushout, we get immediately that $f \amalg f^*$ is in fact an equivalence if $X^B \rightarrow X^A$ is an equivalence. By direct inspection $\mathbf{L}_f C \simeq C$ for any contractible space C (but also notice that a contractible fibrant space is f-local for any f).

B.3 THE CHOICE OF $\lambda = \lambda(A \amalg B)$. The necessary properties of the colimit follow from the correct choice of an ordinal number λ, a choice that depends on the cardinality of $A \amalg B$. In fact we choose an infinite ordinal $\lambda = \lambda(A \amalg B)$ whose cofinality is greater than the cardinality of $A \amalg B$. In the topological case we need to factor maps from the unit interval through a direct limit tower thus we should choose λ to be bigger than the cardinality of I=[0,1], the unit interval. For example, we can choose λ as the first infinite ordinal whose cardinality is greater than that of the product $[0,1] \times (A \amalg B)$ where A, B come from the map f.

In any case our choice of λ is such that guarantees that, for any tower of maps $X_0 \rightarrow X_1 \rightarrow \cdots \rightarrow X_\alpha \rightarrow X_{\alpha+1}$ $(\alpha < \lambda)$ of length λ, there is a canonical equality of function complexes

$$\operatorname{map}(\mathrm{I} \times (A \amalg B), \varinjlim_{\alpha<\lambda} X_\alpha) = \varinjlim_{\alpha<\lambda} \operatorname{map}(\mathrm{I} \times (A \amalg B), X_\alpha).$$

This is the only property of λ that is used in proving the properties of the f-localization.

B.4 HOMOTOPY COMMUTATIVITY OF DIAGRAM (B.1.1). The main technical property of $(\mathbf{L}^\beta X)$ is the homotopy commutativity relation among the maps in (B.1.1) above. Since the tower $\mathbf{L}^\beta X$ is defined inductively from the functor $\mathbf{L}^1 X$ homotopy commutativity of (B.1.1) follows directly from that of the following first step:

(B.4.1)
$$
\begin{array}{ccc}
X^B & \xrightarrow{j^B} & (\mathbf{L}^1 X)^B \\
{\scriptstyle f^*}\downarrow & {\scriptstyle s_1}\nearrow & \downarrow{\scriptstyle f^*} \\
X^A & \xrightarrow{j^A} & (\mathbf{L}^1 X)^A
\end{array}
$$

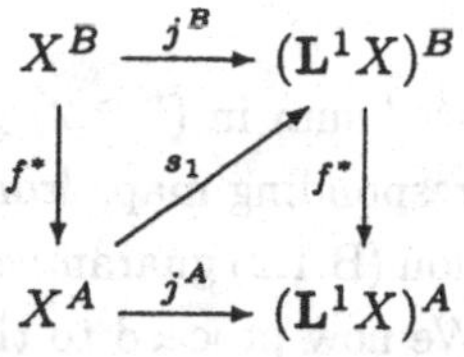

Now the homotopy commutativity follows directly from the homotopy commutativity of the pushout diagram that defines the functor $X \to \mathbf{L}^1_f X = \mathbf{L}^1 X$. Thus since the triangle

(B.4.2)
$$
\begin{array}{ccc}
B \times X^B & \xrightarrow{f^*} & B \times X^A \\
& {\scriptstyle joev}\searrow \quad \swarrow{\scriptstyle s_1'} & \\
& \mathbf{L}^1 X &
\end{array}
$$

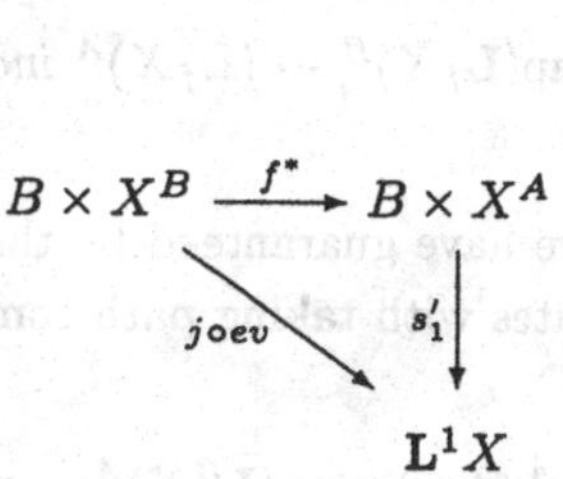

commutes up to homotopy, it follows immediately by adjunction that

(B.4.3)
$$
\begin{array}{ccc}
X^B & \longrightarrow & X^A \\
& \searrow \qquad \downarrow & \\
& (\mathbf{L}^1 X)^B &
\end{array}
$$

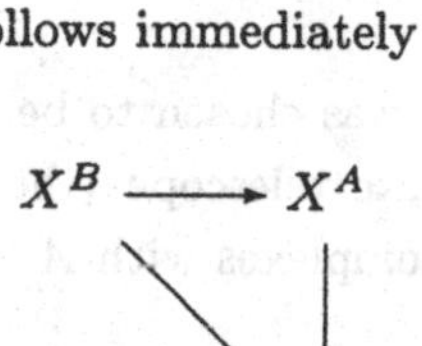

also commutes up to homotopy, and this is precisely the upper left triangle in (B.4.1) By the same reasoning of adjunction, from the homotopy commutative triangle coming from (B.1.2):

(B.4.4)
$$
\begin{array}{ccc}
A \times X^A & \xrightarrow{f \times id} & B \times X^A \\
& {\scriptstyle joev}\searrow \quad \swarrow{\scriptstyle s_1'} & \\
& \mathbf{L}^1 X &
\end{array}
$$

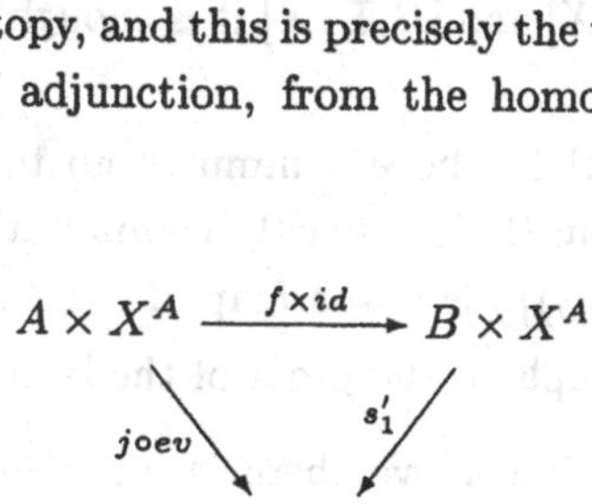

we get the homotopy commutativity of the other triangle (low right) in (B.4.1), namely the triangle:

(B.4.5)
$$\begin{array}{ccc} & X^A & \\ \swarrow & & \searrow{\scriptstyle s_1} \\ (\mathbf{L}^1X)^A & \xleftarrow{f^*} & (\mathbf{L}^1X)^B \end{array}$$

Notice that the amalgamated sum in (B.1.2) guarantees that maps $X^A \longrightarrow (\mathbf{L}^1X)$ are the restriction of corresponding maps from X^B.

To conclude, the construction (B.1.2) guarantees the homotopy commutativity of (B.4.1) and thus of (B.1.1). We now proceed to the main theorem regarding the construction of $\mathbf{L}_fX$.

B.5 THEOREM: *The space* $\mathbf{L}_f^\lambda X = \mathbf{L}^\lambda X$ *defined in (B.1) above is* f*-local.*

B.6 LEMMA: *The natural* $\mathrm{map}(\mathbf{L}_fX)^B \to (\mathbf{L}_fX)^A$ *induces an isomorphism on the set of path components* (π_0).

Proof: Referring to (B.3), we have guaranteed by the choice of λ that taking the direct homotopy limit commutes with taking path components:

$$\pi_0 \operatorname*{colim}_{\beta<\lambda}(\mathbf{L}^\beta X)^A \cong \operatorname*{colim}_{\beta<\lambda} \pi_0(\mathbf{L}^\beta X)^A = \operatorname*{colim}_{\beta<\lambda}[A, \mathbf{L}^\beta X].$$

But by the properties of λ which was chosen to be big enough in comparison with A so that again taking the infinite telescope (the homotopy colimit here) again commutes with taking function complexes with A as the domain, we have also the equation:

$$(\mathbf{L}^\lambda X)^A \equiv (\operatorname*{colim}_{\beta<\lambda} \mathbf{L}^\alpha X)^A = \operatorname*{colim}_{\beta<\lambda}(\mathbf{L}^\beta X)^A.$$

Therefore to consider $[A, \mathbf{L}_fX]$ and $[B, \mathbf{L}_fX]$ it is enough to consider $\operatorname*{colim}_{\beta<\lambda}[A, \mathbf{L}^\beta X]$ etc.

Since the diagram (B.1.1) above commutes up to homotopy, taking π_0 everywhere renders the diagram (B.1.1) strictly commutative and the lemma follows immediately – because $\{\pi_0(s_\beta)\}_\beta$ is now a map of towers of sets; $(\operatorname*{colim}_{\alpha<\lambda} \pi_0 s_\alpha)$ gives an inverse to $\pi_0 f^*$. This completes the proof of the lemma.

Proof of B.5: By Lemma B.6 above there is an isomorphism between the sets of path components of $\mathrm{map}(A, \mathbf{L}^\lambda X)$ and $\mathrm{map}(B, \mathbf{L}^\lambda X)$. We must show that any two corresponding path components are weakly homotopy equivalent, namely the natural map

$$f^* : (\mathbf{L}^\lambda X)^B_\varphi \longrightarrow (\mathbf{L}^\lambda X)^A_{\varphi \circ f}$$

induces an isomorphism on the homotopy groups $\pi_i(\ ,*)$ with the proper choice of base points. Notice that the map f^* is the limit of a map of towers: the towers $(\mathbf{L}^\beta X)^B_{\beta \geq 0}$ and $(\mathbf{L}^\beta X)^A_{\beta \geq 0}$.

These maps induced by f by composition are all denoted here by f^* and they form a map of towers, since the diagram

$$\begin{array}{ccc} (\mathbf{L}^\beta X)^B & \xrightarrow{f^*} & (\mathbf{L}^\beta X)^A \\ \downarrow & & \downarrow \\ (\mathbf{L}^{\beta+1} X)^B & \xrightarrow{f^*} & (\mathbf{L}^{\beta+1} X)^A \end{array}$$

commutes strictly. The main technical difficulty of the proof is that the collection of maps in (B.1.1) s^*_β in the other direction does not form a map of towers since the corresponding ladder, namely the square in (B.1.1), commutes only up to homotopy. Thus $\operatorname{colim} s^*_\beta$ is *not* a well defined map. It is possible, with some effort, to modify $(\mathbf{L}^\beta X)_\beta$ in such a way as to make the corresponding maps s^*_α into maps of towers that do give a map backward, $(\mathbf{L}^\lambda X)^A \to (\mathbf{L}^\lambda X)^B$, which is a homotopy inverse to $\operatorname{map}(f, \mathbf{L}^\lambda X)$ [H].

To get our result more quickly here we only need to show that $\operatorname{map}(f, \mathbf{L}^\lambda X)$ is a weak equivalence, so we get a map backward only on the level of homotopy groups, i.e. a map of groups

(B.6.1) $$\pi_*(\mathbf{L}^\lambda X)^A_{\varphi \circ f} \to \pi_*(\mathbf{L}^\lambda X)^B_\varphi$$

where the subscript denotes pointed connected components for some map $\varphi : B \to \mathbf{L}^\lambda X$. The main technical difficulty in obtaining this map stems from the fact that we are dealing with both non-connected and unpointed mapping spaces: Even if we had base points in the given spaces X, A, B the two typical components of the mapping spaces we need to compare are unpointed and the diagram (B.1.1) as it stands does not commute up to base point preserving homotopies.

To get around difficulties, first notice that by Lemma B.6 proven above, $\operatorname{map}(f, \mathbf{L}^\lambda X)$ induces an isomorphism on the set of path components π_0. So for each corresponding pair of components we choose base points in $(\mathbf{L}^\lambda X)^A$ and $(\mathbf{L}^\lambda X)^B$ and we must show that, with respect to these base points, our map induces an isomorphism on the homotopy groups of these two corresponding components.

INTEGRAL HOMOLOGY ISOMORPHISM We notice that by the same argument as in Lemma B.6 we can conclude:

B.6.2 LEMMA: *The natural map* $(\mathbf{L}_f X)^B \to (\mathbf{L}_f X)^A$ *induces an isomorphism on the integral homology groups* $H_*(-, \mathbb{Z})$.

Proof: Integral homology commutes with direct linear limits just as the set of path component. Similarly applying homology to homotopy commutative diagram B.1.1 renders it strictly commutative. Therefore again we $\operatorname{colim}_\beta H_*(s_\alpha, \mathbb{Z})$ is an inverse to $H_*(f^*, \mathbb{Z})$.

To complete the proof we now notice that each path component of any mapping space $\operatorname{map}_*(W, T)$ for W a cofibrant space is $H\mathbb{Z}$-local in the sense of Bousfield [B-1], see (E.4) below, since (using elementary fact (1.A.8)(e.3)) it is a homotopy inverse limit of a tower of nilpotent spaces which are local with respect to any map that induces an isomorphism on integral homology: It is easy to see by induction the well known fact that the function complex $\operatorname{map}_*(W, T)$ is nilpotent if W is finite dimensional. Therefore the map f^* when restricted to each corresponding pair of path components is an integral homology isomorphism between two spaces that are local with respect to (maps that induce isomorphism on) integral homology thus it is a weak equivalence and so our space $\mathbf{L}_f^\lambda X$ is f-local. ∎

Of course we have just used the theory of homological localization [B-1] together with elementary properties of $H\mathbb{Z}$-local spaces to construct general homotopical localization. This is not the most elementary way to develop general localization theory! Therefore we give below alternative treatments: one of them more elementary, the other more direct. First the direct method.

UNPOINTED MAPS OF SPHERES Rather then concluding the proof of (A.5) using integral homology and HZ-local spaces one can proceed using a recent theorem of Casacuberta and Rodriguez [Ca-R] that asserts that a map between two pointed function complexes $\operatorname{map}_*(B, X) \longrightarrow \operatorname{map}_*(A, X)$ induced by f is an equivalence if this map induces an isomorphism on unpointed homotopy classes from the unpointed spheres. Since one can restrict to any component of X taking the homotopy fibres of the evaluation maps to the spaces in the tower $\mathbf{L}^\beta\mathbf{X}$ in diagram (B.1.1) one gets the same diagram but with pointed mapping spaces rather than the unpointed ones in (B.1.1). Now that diagram of pointed function complexes is also homotopy commutative. Therefore it induces in the direct limit isomorphisms on the unpointed homotopy classes of unpointed maps from any space – since we are using *finite* spaces to map in, we still get the analog of Lemma A.6 above for S^n, the unpointed sphere rather than S^0. Using [Ca-R] we deduce that our map $\operatorname{map}_*(f, \mathbf{L}^\beta X)$ is an equivalence. Therefore the corresponding unpointed mapping spaces are also equivalent as needed. The proof of the result in [Ca-R] uses the fact from group theory that a map of nilpotent groups that induces surjection on conjugacy classes is a surjective and the fact that the above spaces of pointed maps are always homotopy inverse limits of nilpotent spaces.

ALTERNATIVE, ELEMENTARY, PROOF OF A.5 To get a self-contained proof we make some preparations (arguing simplicially to ensure that inclusions are cofibrations). We would like to get the diagram (B.1.1) to be homotopy commutative in the pointed category so that it would imply that (B.6.1) exists and is an isomorphism. Let us choose a representative map $g : B \to \mathbf{L}^\lambda X$. Without loss of generality we can assume that g is a limit of maps g_β: $B \to \mathbf{L}^\beta X$ for $\beta \geq 0$, if g_β does not pull back to a map to X but rather to some g^α: $B \longrightarrow \mathbf{L}^\alpha X$, then modify $\mathbf{L}^\beta X$ only for $\beta \geq \alpha$.

For each such a choice of g, g_β we now modify slightly the spaces $\mathbf{L}^\beta X$ to get $\tilde{\mathbf{L}}^\beta X$. These will be quotient spaces of the same homotopy type, with the property that they are all pointed; moreover they all come with pointed maps $B \to \tilde{\mathbf{L}}^\beta X$, and pointed $\tilde{s}_\beta$: $(\tilde{\mathbf{L}}^\beta X)^A \to (\tilde{\mathbf{L}}^\beta X)^B$. These base point will allow us to compare the homotopy groups of corresponding components. In order to render (B.1.1) homotopy commutative in the pointed category we render (B.4.1) and (B.4.4) homotopy commutative relative to our chosen copy of B and A respectively. The diagram corresponding to (B.1.1) above will now consist of base point preserving maps and will be homotopy commutative in $\mathcal{S}_*$. Therefore it gives rise to a corresponding strictly commutative diagram of the corresponding homotopy groups.

We now define $\tilde{\mathbf{L}}^\beta X$ for a given $g^\lambda = g$: $B \to \mathbf{L}^\lambda X$ for which we have assumed without loss of generality that it pulls back to g^0: $B \to X$. We define $\tilde{\mathbf{L}}^1 X$ to be the quotient of $\mathbf{L}^1 X$ obtained as follows: In (B.1.2) we first turn, in the canonical way, the map $f \amalg f^*$ into a cofibration $c(f \amalg f^*)$ by replacing the range $B \times X^A$ by the mapping cylinder $\mathrm{Cyl}(f \amalg f^*)$.

Second and crucially, in the resulting mapping cylinder we collapse a copy of $\Delta[1] \times B \times g^0$ down to B itself. Notice that this cylinder over B is embedded in $\mathrm{Cyl}(f \amalg f^*)$ since the top map in (B.1.2) is an isomorphism when restricted to $B \times g^0$. This does not change the homotopy type of the spaces involved.

Finally we take $\tilde{\mathbf{L}}^1 X$ to be the <u>strict</u> pushout along the resulting top cofibration $c(f \amalg f^*)$, namely, further to step two, we identify points in the mapping cylinder that correspond along the evaluation map ev.

Since the top map in (B.1.2), $f \amalg f^*$, and ev are surjective one gets a certain identification space of:

$$\Delta[1] \times (A \times X^A \coprod_{A \times X^B} B \times X^B).$$

This collapsing ensures that the two maps of B to the double cylinder $\tilde{\mathbf{L}}^1X$, the one coming from g^0: $B \to X$ and the other from the point in $B \times X^A$, namely $\{*\} \times f \circ g^0$, are identical and the homotopies given by the homotopy commutative triangles (B.4.2, B.4.4) above are now pointed homotopies that preserve this map that is now taken as a base point in the mapping spaces involved.

Now that we have pointed maps and homotopies we can apply homotopy groups to the diagram (B.1.1) above (after changing L to $\tilde{L}$ everywhere) for the appropriate components of the mapping spaces. Since homotopy groups commute with these linear colimits, we get immediately that $\pi_* f^*$ and $\pi_* s_\lambda^*$ are mutual inverses. Thus f^* induces a weak equivalence as needed. This completes the proof of (B.5). ∎

B.7 Claim: *Let P be an f-local space. Then for each ordinal $\beta \geq 0$ the natural map induced by coaugmentation*

$$\text{map}(\mathbf{L}_f^\beta X, P) \to \text{map}(X, P)$$

is a homotopy equivalence.

Proof: Notice that the tower $(\mathbf{L}_f^\beta X$ in (B.1) is a tower of cofibrations by our definition, so its homotopy colimit is equivalent to its direct limit. We use the basic fact (Appendix HL) that the mapping space functor: map$(-, P)$ turns a homotopy direct limit into a homotopy inverse limit. Therefore upon applying this functor to the tower of spaces $\mathbf{L}^\beta X$ we see that for any limit ordinal β the map in claim B.7 is a map between two homotopy (inverse) limits

$$holim_\alpha \text{map}(\mathbf{L}_f^\alpha X, P) \to holim_\alpha \text{map}(X, P) = \text{map}(X, P).$$

Again it follows from the basic properties of homotopy limits that if for each α the above map is an equivalence, then so is the induced map on the homotopy limits. It follows that it is sufficient to prove the claim for non-limit ordinals and by the inductive definition therefore it is sufficient to consider the case $\mathbf{L}_f^1 X$ for an arbitrary space X. Referring to the definition of $\mathbf{L}_f^1 X$ in (B.1.2) above as a homotopy pushout, one again uses the observation that taking function complexes with P as a range turns any homotopy pushout square into a homotopy pullback square. Because in such squares pulling along a weak equivalence gives a weak equivalence we only need to verify that the map:

$$\text{map}(C, P) \to \text{map}(B \times X^A, P)$$

where $C = A \times X^A \coprod_{A \times X^B} B \times X^B$ is the space in the left hand corner of (B.1.2), is a homotopy equivalence. Once again we use the fact that C is the homotopy pushout by definition. But it is immediate from the assumption on P that $\operatorname{map}(B, P) \simeq \operatorname{map}(A, P)$ is a homotopy equivalence. Then notice that by the exponential law for function complexes, it follows from the assumption on P that $\operatorname{map}(B \times X^W, P) \simeq (A \times X^W, P)$ for any (cofibrant) space W. Now C being a homotopy pushout, we show that $\operatorname{map}(C, P) \xrightarrow{\cong} \operatorname{map}(A \times X^A, P)$ is a homotopy equivalence, and thus so is $\operatorname{map}(B \times X^A, P) \to \operatorname{map}(C, P)$. To accomplish this final step consider the diagram that presents C as the homotopy pushout with the well defined map to $B \times X^A$ that follows from the strict commutativity of the square (appendix [HL]):

(B.7.1)

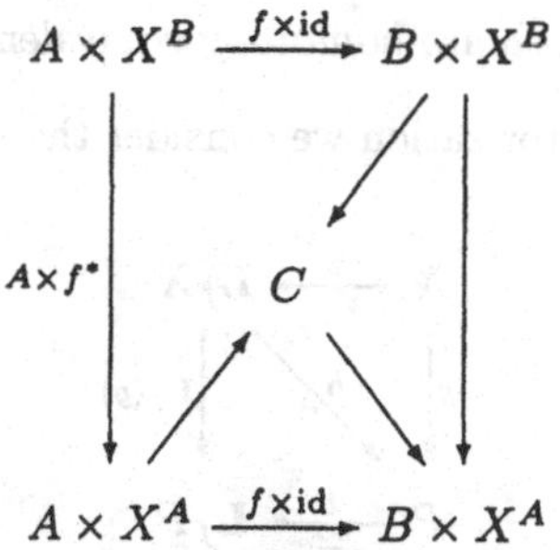

The diagram is commutative, since C by definition is the (homotopy) pushout.

Since both maps $f \times \mathrm{id}$ induce homotopy equivalences: map $(f \times \mathrm{id}, P)$, we get by elementary pullback that $\operatorname{map}(A \times f^*, P)$ is a homotopy equivalence and thus so are the maps $\operatorname{map}(B \times X^A, P) \longrightarrow \operatorname{map}(C, P) \longrightarrow \operatorname{map}(A \times X^A, P)$. This completes the proof of (B.7). ∎

We now address the claim that $\mathbf{L}_f$ is idempotent. In fact we saw above (B.5) that $\mathbf{L}_f X$ is always f-local, so we combine this with (B.7) and address idempotency and universality of $\mathbf{L}_f$.

C. Universality and continuity of $\mathbf{L}_f$

We now collect the results of Section 1.B above and conclude the proof of Theorem A.3 about the existence of the f-localization functor $\mathbf{L}_f$. This leads naturally to a discussion of the universality properties of $\mathbf{L}_f$ or, more correctly, universality properties of the coaugmentation $X \to \mathbf{L}_f X$.

The discussion above leads to the characterization of the functor $\mathbf{L}_f$ in terms of three basic properties.

(i) The space $\mathbf{L}_f X$ is f-local.

(ii) The coaugmentation map $Y \longrightarrow \mathbf{L}_f Y$ is a weak equivalence when applied to an f-local space Y.

(iii) For any f-local space P the map induced by the coaugmentation a, namely map(a, P), is a weak equivalence.

C.1 THEOREM: *For any map $X \to P$ into an f-local space there exists a factorization $X \to L_f X \to P$ which is unique up to homotopy.*

Remark: The above proposition says that $X \to L_f X$ is initial among all maps of X to f-local spaces. Below we will see that the same map is also a terminal map in a certain class of maps out of X called f-local equivalences, or L_f equivalences.

C.2 COROLLARY: *Consider the two maps $u, v : L_f X \to L_f L_f X$ where $u = L_f(a)$ and $v = a_{L_f}$ (i.e., u is L_f applied to the coaugmentation and v is the coaugmentation on $L_f X$). Then $v \sim u$ and both are homotopy equivalences.*

Proof of C.1: To get the factorization we consider the square

$$
\begin{array}{ccc}
X & \xrightarrow{\ a\ } & L_f X \\
{\scriptstyle g}\downarrow & {\scriptstyle h}\nearrow & \downarrow{\scriptstyle L_f(g)} \\
P & \xrightarrow[\simeq]{\ b\ } & L_f P
\end{array}
$$

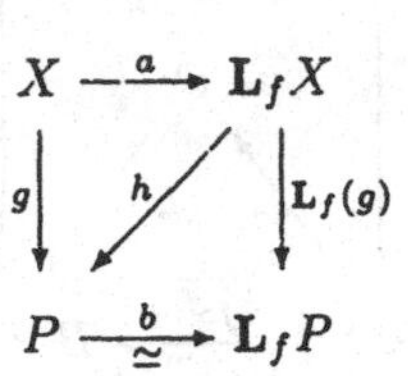

We know from (B.2) above that for the f-local space P the coaugmentation b is a weak homotopy equivalence.

Now since P is local, we saw in (B.7) above that

$$\mathrm{map}(L_f X, P) \to \mathrm{map}(X, P)$$

is a homotopy equivalence. Looking at the induces isomorphism on the set of path components we see that there is exactly one homotopy class $L_f X \to P$ that pulls back to the given map $X \to P$. ∎

Before deriving the corollary we formulate a general observation that will be used repeatedly:

C.3 PROPOSITION (DIVISIBILITY): *For any two maps $g, h : L_f X \overset{h}{\underset{g}{\rightrightarrows}} P$ into any f-local space P, one has $h \sim g$ if and only if the compositions $h \circ a \sim g \circ a$ with the coaugmentation $a : X \to L_f X$ are homotopic.*

Proof: As above this follows immediately from the equivalence of function complexes (B.7) above.

Proof of Corollary C.2: Both maps pull back to the same map $X \to \mathbf{L}_f\mathbf{L}_f X$ in the commutative square:

$$\begin{array}{ccc} X & \xrightarrow{a} & \mathbf{L}_f X \\ {\scriptstyle a}\downarrow & & \downarrow{\scriptstyle a'} \\ \mathbf{L}_f X & \xrightarrow{\mathbf{L}_f(a)} & \mathbf{L}_f\mathbf{L}_f X \end{array}$$

Thus by divisibility we get the desired homotopy $v \sim u$. We saw above that $a_{\mathbf{L}_f} = v$ is a homotopy equivalence, thus so is u.

Notice a further useful consequence of universality (idempotency)

C.4 PROPOSITION (NO ZERO DIVISORS): *Let W be a retract of $\mathbf{L}_f X$. If the composition $X \to \mathbf{L}_f X \to W$ is null homotopic then $W \simeq *$. In particular $\mathbf{L}_f(*) \simeq *$ and $\mathbf{L}_f$(nullmap) is a null map.*

Proof: First notice that by (B.2) above $\mathbf{L}_f$ turns a contractible space into a contractible one. Since a null map factors through a contractible space $\mathbf{L}_f$ turns a null-homotopic map into a null-homotopic map. The first claims then follow immediately by applying $\mathbf{L}_f$ to the composition. Since W is a retract of a local space it is local. By idempotency (C.2) we get that $\mathbf{L}_f(X \to \mathbf{L}_f X \to W)$ is just the retraction map $\mathbf{L}_f X \to W$, so the retraction is null homotopic since the composition before applying localization – and thus also after – is null. Thus W is contractible.

UNIVERSALITY CONTINUED, f-LOCAL EQUIVALENCE. The discussions in (C.1) and (C.2) above prove the universal properties of $\mathbf{L}_f$ claimed in (A.3). The proof that the above construction leads to a continuous (or simplicial) functor $\mathbf{L}_f$ is given in (C.11) below.

However, the functor $\mathbf{L}_f$ is, like any other coaugmented idempotent functor, universal also in another sense: The coaugmentation map is also terminal among a certain class of maps out of X. This class is interesting for its own sake, namely the class of f-local equivalences, sometimes called $\mathbf{L}_f$-equivalence:

C.5 DEFINITION AND PROPOSITION: *A map $g : X \to Y$ is called an f-local equivalence or $\mathbf{L}_f$-equivalence if it satisfies one of the two equivalent conditions:*

(i) *For all f-local T the induced map* $\mathrm{map}(g, T)$ *is an equivalence.*
(ii) *The map g induces a homotopy equivalence $\mathbf{L}_f(g) : \mathbf{L}_f X \to \mathbf{L}_f Y$.*

In this definition, if X or Y are not cofibrant one should first replace them by weakly equivalent cofibrant spaces.

Proof: This is an easy consequence of (B.7), (C.1) and (C.2) above.

C.5.1 Example: Of course the principal example of an f-local equivalence is the coaugmentation map, see (B.7) above.

C.6 Proposition: *For any map $g : X \to Y$ which is an $\mathbf{L}_f$-equivalence there is an extension $X \to Y \xrightarrow{\ell} \mathbf{L}_f X$ that is unique up to homotopy with $\ell \cdot g \sim a : X \to \mathbf{L}_f X$.*

Proof: We get ℓ by using the homotopy inverse to $\mathbf{L}_f(g)$. Uniqueness of $[\ell]$ follows from divisibility above (C.3).

C.7 Proposition: *($\mathbf{L}_f$ is terminal) The coaugmentation map $X \to \mathbf{L}_f X$ is terminal, up to homotopy, among all f-local equivalences $e : X \to Y$: For any such e there exists a map $g : Y \to \mathbf{L}_f X$, unique up to homotopy, such that $g \circ e$ is homotopic to the coaugmentation a.*

Proof: This follows directly from (C.6) and the fact (C.5.1) that the coaugmentation is always an f-local equivalence.

Continuity of $\mathbf{L}_f$: A convenient property of the functor $\mathbf{L}_f$ is 'continuity'. This means naively that the set-function given by functoriality:

$$\operatorname{map}(X, Y) \to \operatorname{map}(\mathbf{L}_f X, \mathbf{L}_f Y)$$

is a continuous map when the usual 'compactly generated' topology is assigned to these sets of maps. Here we find it more convenient to regard our category of spaces $\mathcal{S}$ as a simplicial category with a morphism set $\operatorname{map}(X, Y)$ a simplicial set whose n-simplices are maps $\Delta[n] \times X \to Y$. From this point of view a continuous functor is just a *simplicial functor*, namely one which preserves the simplicial structure given on hom-sets: So it induces a simplicial map $\operatorname{map}(X, Y) \to \operatorname{map}(FX, FY)$, natural and respecting composition. Even the very existence of a simplicial map is not guaranteed by the functorial properties of F – see examples in (C.9.1).

Given a simplex in $\operatorname{map}(X, Y)$, namely a map $\xi : \Delta[n] \times X \to Y$, we want to assign to it a map $\Delta[n] \times \mathbf{L}_f X \to \mathbf{L}_f Y$ in $\operatorname{map}(\mathbf{L}_f X, \mathbf{L}_f Y)$. One first considers $\mathbf{L}_f(\xi) : \mathbf{L}_f(\Delta[n] \times X) \to \mathbf{L}_f Y$ but then one needs a map:

$$\Delta[n] \times \mathbf{L}_f X \to \mathbf{L}_f(\Delta[n] \times X).$$

This discussion motivates our definition in the category of simplicial sets: A similar definition can obtain in the topological category with simplicial function complexes.

C.8 DEFINITION: *A coaugmented functor* $\mathrm{Id} \xrightarrow{j} \mathbf{F} : \mathcal{S} \to \mathcal{S}$ *will be called simplicial if for all* $X, K \in \mathcal{S}$ *there exists a map:*

$$\sigma : K \times \mathbf{F}X \to \mathbf{F}(K \times X)$$

which is natural in both K *and* X.

and, in addition, satisfies the following:

(i) If $*$ *is the one-point space then the following composition is the identity:*

$$\mathbf{F}X \xrightarrow{\cong} * \times \mathbf{F}X \xrightarrow{\sigma} \mathbf{F}(X \times *) \xrightarrow{\cong} \mathbf{F}X$$

(ii) It renders the following natural diagram commutative for any X, K, L:

$$\begin{array}{ccc}
K \times X & \xrightarrow{j} & \mathbf{F}(K \times X) \\
{\scriptstyle K\times j}\downarrow & \overset{\sigma}{\nearrow} & \downarrow \\
K \times \mathbf{F}X & \longrightarrow & \mathbf{F}(K \times \mathbf{F}X)
\end{array}$$

(iii) It renders the following natural diagram commutative for any X, K, L:

$$\begin{array}{ccc}
K \times L \times \mathbf{F}X & \xrightarrow{K\times\sigma} & K \times \mathbf{F}(L \times X) \\
 & \underset{\sigma}{\searrow} & \downarrow{\scriptstyle\sigma} \\
 & & \mathbf{F}(K \times L \times X)
\end{array}$$

C.9 PROPOSITION: *Any simplicial functor* $\mathbf{F}$ *preserves the simplicial structure of* $\mathcal{S}$, *inducing a map as above between function complexes. Further, this map is natural in both variables and respects compositions.*

Proof: (Compare [B-3], [HH]) This is a straightforward consequence of the naturality of σ in both variables and properties (i)–(iii) that are designed precisely to render the necessary diagrams commutative.

C.9.1 A COUNTER-EXAMPLE TO CONTINUITY The coaugmented functor $X \to \overline{W}GX$ does not preserve the simplicial structure, even though here as often it can be slightly modified to preserve that structure [D-K-1], [Q-1]. For another example, let A be any space and define a functor $\mathbf{W}_A = \mathbf{W} : \mathcal{S}_* \to \mathcal{S}_*$ as a large wedge;

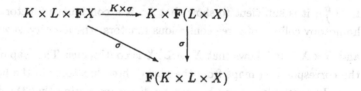

$\mathbf{W}_A X = \mathbf{W}X = \bigvee_{A \to X} A$, namely we take one copy of A for each $g : A \to X$. This is certainly functorial as the copy corresponding to g (as above) maps by identity under $f : X \to Y$ to the copy corresponding to $f \circ g$. But $\mathbf{W}$ is *not* continuous. If we take $A = S^o$ and $X = S^o$ $Y = I$, the unit interval, then $\mathrm{map}(S^o, I) = I$ and $\mathbf{W}S^o = S^o$ while $\mathbf{W}I = I^\delta$ (I^δ is I with the discrete topology). The function $\mathrm{map}_*(S^0, I) \to \mathrm{map}(\mathbf{W}S^o, \mathbf{W}I) = \mathrm{map}_*(S^o, I^\delta)$ in not continuous. Of course the above counter-examples are functors that do not preserve weak equivalences. A functor that preserves weak equivalences has been shown to be 'essentially' continuous by the work of Dwyer and Kan about computing function complexes using only the model category structure via simplicial localizations: [D-K-1].

C.9.2 Example: Given any space $T \in \mathcal{S}$ the functor $\mathrm{map}(T, -)$ is simplicial: $\sigma(k, f) = k \times f$ where $k : T \to K$ is the constant function $(\Delta[n] \times T \xrightarrow{pr_1} \Delta[n] \xrightarrow{k} K)$.

C.10 Proposition: *Let $\underset{\sim}{F}$ be any I-diagram of functors and natural transformations between them. If for each $\alpha \in I$ the functor $F_\alpha : \mathcal{S} \to \mathcal{S}$ is simplicial, then so* is $\operatorname{hocolim}_I \underset{\sim}{F}$.

Proof: This follows from naturality of the defining map σ (C.8). The map σ is assembled from the corresponding maps on each space in the diagram using the naturality.

C.11 Continuity of $\mathbf{L}_f$. We now address the question of continuity of the functors $\mathbf{L}_f$ for $f : A \to B$. Since $\mathbf{L}_f$ is the (homotopy) colimit of $(\mathbf{L}_f^\beta)_{\beta < \lambda}$ and $\mathbf{L}_f^{\beta+1} = \mathbf{L}_f^1(\mathbf{L}_f^\beta)$, it is sufficient to show that $\mathbf{L}_f^1$ is a continuous functor. But $\mathbf{L}_f^1$ itself is a homotopy colimit of three continuous functors: the identity, $A \times X^A \underset{A \times X^B}{\amalg} B \times X^B$ and $B \times X^A$; it follows that $X \to \mathbf{L}_f^1 X$ is continuous: The map σ is assembled from the corresponding maps from the above three functors via the hocolim functor.

In particular, it can be seen by direct inspection that the diagram:

(C.12)
$$\begin{array}{ccc} K \times X & \longrightarrow & K \times \mathbf{L}_f^\beta X \\ \downarrow & \overset{\sigma}{\swarrow} & \downarrow \\ \mathbf{L}_f^\beta(K \times X) & \longrightarrow & \mathbf{L}_f^\beta(K \times \mathbf{L}_f^\beta X) \end{array}$$

is commutative for $\beta = 1$, it follows that it is so for all $\beta \leq \lambda$. The same argument works for the other conditions of (C.8). Thus we can conclude:

C.13 THEOREM: *The functor* $\mathbf{L}_f$ *is simplicial (C.8).* ∎

D. $\mathbf{L}_f$ and homotopy colimits, f-local equivalence

In this section we analyze some relations between $\mathbf{L}_f$ and homotopy colimits (i.e. homotopy direct limits) both pointed and unpointed. We would like to find a relation between the localization of the homotopy colimit of a commutative diagram of spaces and the homotopy colimit of the localizations of individual spaces in the diagram. While in general localization does not commute with homotopy colimits, these colimits preserve f-local equivalence (D.2) and (D.3) below. In a special case nullification with respect to a finite space W does commutes with linear homotopy colimits (Telescopes), see (D.6) below.

In Appendix HL the reader can find a short account of some relevant material about homotopy (co)limits. In fact we will work here mostly in S_* and S_*^I, but everything we say holds with the obvious changes in S and S^I.

D.1 SIMPLICIAL AND DISCRETE DIAGRAMS In general we will work with a diagram over a small category I. But sometimes it is more natural and useful to consider a diagram over a simplicial category, i.e. a category whose morphism sets are enriched by a simplicial structure [Mac], for example, when considering a topological or simplicial group acting on a space. Here we simply notice that homotopy colimits over a (small) simplicial category $\mathcal{G}$, i.e. a simplicial functor $\underset{\sim}{X} : \mathcal{G} \to S_*$, can be expressed as a composition of homotopy colimits over set-valued categories; see for example [D-K-1]. One simply takes first the homotopy colimits over the small categories $\mathcal{G}_n$ and then puts them all together by realizing the resulting simplicial space.

The main results of this section are:

D.2 PROPOSITION: *Let* $g : \underset{\sim}{X} \to \underset{\sim}{Y}$ *be a map of (pointed) I-diagrams of cofibrant spaces in S. Assume that* $g_\alpha : X_\alpha \to Y_\alpha$ *is an* $\mathbf{L}_f$*-equivalence for* $f : A \to B$*. Then hocolim* $g : \operatorname{hocolim}_I \underset{\sim}{X} \to \operatorname{hocolim}_I \underset{\sim}{Y}$ *is also an* $\mathbf{L}_f$*-equivalence. (The same is true for pointed* $\operatorname{hocolim}_*$*.)*

D.2.1 REMARK : One of the main examples of Proposition D.2 above is the localization map on any diagram of spaces $\underset{\sim}{X} \to \mathbf{L}_f\underset{\sim}{X}$. Here we use the fact that $\mathbf{L}_f$ is defined as a functor on the category of spaces and not just on the associated homotopy category. One gets a map of commutative diagrams such that, for each $\alpha \in I$, the map is an f-local equivalence. Now using the equivalence of the two conditions that define f-local equivalence we get as a corollary the following commutation rule for localizing homotopy colimits where the commutation map claimed can be constructed using the simplicial structure on $\mathbf{L}_f$:

D.3 THEOREM: *Given a small category I and a diagram over it $\underset{\sim}{X} : I \to \mathcal{S}_*$ of cofibrant space, the natural map obtained by applying localization to the coaugmentation: $\mathbf{L}_f(a) : \mathbf{L}_f \operatorname{hocolim}_{*I} \underset{\sim}{X} \to \mathbf{L}_f \operatorname{hocolim}_{*I} \mathbf{L}_f \underset{\sim}{X}$, is a weak homotopy equivalence. Moreover, there is a natural map c (comparison or commutation map) $c : \operatorname{hocolim}_{*I} \mathbf{L}_f \underset{\sim}{X} \to \mathbf{L}_f \operatorname{hocolim}_{*I} \underset{\sim}{X}$ such that $\mathbf{L}_f(c)$ is an inverse to the above $\mathbf{L}_f(a)$ and thus it is a homotopy equivalence. (The same is true for unpointed $\operatorname{hocolim}_I$.)* ∎

Proof of (D.2): This follows immediately from writing the function complex out of a hocolim as a homotopy inverse limit: Let P be any f-local space; we must show by Definition (C.5) of $\mathbf{L}_f$-equivalence above that the induced map $\operatorname{map}_*(\operatorname{hocolim}_I g, P)$: $\operatorname{map}_*(\operatorname{hocolim}_I \underset{\sim}{Y}, P) \to \operatorname{map}_*(\operatorname{hocolim}_I \underset{\sim}{X}, P)$ is a weak equivalence. But by the fundamental property of hocolim it follows that $\operatorname{map}_*(\operatorname{hocolim}_I \underset{\sim}{X}, P)$ is weakly equivalent to the homotopy inverse limit of function complexes: $\operatorname{holim} \operatorname{map}_*(\underset{\sim}{X}, P)$, where the homotopy limits and colimits are taken over the categories I and I^{op} see Appendix HL below. Now by assumption, for each $\alpha \in I$ the induced map $\operatorname{map}_*(g_\alpha, P)$ is a weak equivalence, therefore taking homotopy limits we still get a weak equivalence as needed.

D.4 EXAMPLE: A basic example of $\mathbf{L}_f$-equivalence is E_*-equivalence for some generalized homology theory E_*. In this case, in order to express this equivalence as an $\mathbf{L}_f$-equivalence we take f to be the (pointed)-union of all pointed E_*-equivalences $A_\alpha \to B_\alpha$ with cardinality $A_\alpha \amalg B_\alpha$ limited by $E_*(pt)$: we can take the first infinite cardinality bigger than $E_*(pt)$.

D.5 EXAMPLE: If $g : X \to Y$ is any (pointed) $\mathbf{L}_f$-equivalence of cofibrant spaces then it is not hard to check from the general result above about homotopy colimits that so are the following:

(i) $W \times X : W \times X \to W \times Y$.

(ii) $\Sigma g : \Sigma X \to \Sigma Y$.

(iii $W \wedge g : W \wedge X \to W \wedge Y$ (for a pointed W).

(iv) For any pointed spaces X, Y one has $\mathbf{L}_f(X \vee Y) \simeq \mathbf{L}_f\big(\mathbf{L}_f X \vee \mathbf{L}_f Y\big)$. The same goes for smash products of two pointed spaces.

(v) In any cofibration sequence $A \longrightarrow X \xrightarrow{g} X/A$, if $\mathbf{L}_f A \simeq *$ then $\mathbf{L}_f(g)$ is a weak equivalence.

Under special but still useful assumptions, taking homotopy colimit does commute with taking localization .

D.6 PROPOSITION: *Let $X_0 \to X_1 \to X_2 \to \cdots \to X_\infty$ be a telescope of cofibrations with X_∞ the infinite telescope or homotopy colimit of the tower. If W is a finite space then the natural map* $\operatorname{hocolim}_i \mathbf{P}_W(X_i) \xrightarrow{\simeq} \mathbf{P}_W X_\infty$ *is an equivalence.*

Proof: (Arguing simplicially) Using (D.3) it is enough to show that the left hand side is W-null. But W being a finite space, taking function complex out of W commutes with taking infinite telescopes: $\operatorname{map}_*(W, Y_\infty) \simeq \operatorname{hocolim}_i \operatorname{map}_*(W, Y_i)$ for any tower (Y_i). We can conclude that $\operatorname{map}_*(W, \operatorname{hocolim} \mathbf{P}_W(X_i))$ is given as a homotopy colimit of contractible spaces hence it is contractible as needed.

D.7 EXAMPLE As a final example of an f-equivalence consider the following. If A is a space such that $\mathbf{L}_f A \simeq *$ for some map f then for any X the nullification map $X \longrightarrow \mathbf{P}_A X$ is an f-equivalence. This can be seen either directly from the construction of $\mathbf{P}_A$ via a transfinite telescope where the assumption leads directly to the result that each step in it is an f-equivalence, or using for example Theorems H.1 and H.2 below. Since the fibre F of $X \longrightarrow \mathbf{P}_A X$ is killed by $\mathbf{P}_A$ and $\mathbf{L}_f$ kills A, $\mathbf{L}_f$ kills also that fibre F. therefore by (H.1) the map $X \longrightarrow \mathbf{P}_A X$ becomes an equivalence upon applying $\mathbf{L}_f$ to it as claimed. the map

D.8 EXAMPLE: f-EQUIVALENCE IN FIBRATIONS We saw above in (D.5) that sometimes f-equivalence respects cofibration. Here is an example where it interacts with fibrations. This will shed light on fibrewise localization in Section F below.

Claim. Given any principal fibrations ladder over base spaces $B, \overline{B}$:

(D.8.1)
$$\begin{array}{ccccc} G & \longrightarrow & E & \longrightarrow & B \\ \downarrow g & & \downarrow h & & \downarrow k \\ \overline{G} & \longrightarrow & \overline{E} & \longrightarrow & \overline{B} \end{array}$$

If both g and h are f-equivalences then so the map k. The point is that for a principal fibration one can construct the base space as the bar-construction: It is a homotopy colimit of a diagram consisting of various products of the group and the total space: $B \simeq \mathbf{B}(G, E) \equiv \|(G \times G \cdots \times G \times E\|$, where $\| - -\|$ denotes the realization of simplicial space, an important example of homotopy colimit [S], Appendix HC. It now follows from (D.5(i)) and (D.3) above that $\mathbf{B}(G, E) \longrightarrow \mathbf{B}(\overline{G}, \overline{E})$ is an f-equivalence as claimed since it is such before taking realizations.

As a corollary we get another comparison: If (D.8.1) is now a ladder of general fibrations and with equal connected bases $B = \overline{B}$, and if g is an f-equivalence then so is h. This follows easily by backing up the fibrations to get principal fibrations with equal groups: $\Omega B = \Omega \overline{B}$.

The above claim is definitely not true for general non-principal fibration sequences. In particular it is possible for both g and h above to be homology isomorphisms without k being so.

D.9 Counter-example: On the other hand, the homotopy inverse limits of weakly $\mathbf{L}_f$-equivalent diagrams $\underset{\sim}{X} \to \underset{\sim}{Y}$ most often are *not* $\mathbf{L}_f$-equivalent. Consider the fact that if one applies the loop space functor to a map that is an E_*-equivalence for some homology theory, or even an $H\mathbb{Z}$-equivalence, then one will not in general get an E_*-equivalence.

E. Examples: Localization according to Quillen–Sullivan, Bousfield–Kan, homological localizations, and v_1-periodic localization

In this section we present some of the more useful examples of localization functors. They all depend on the proper choice of a map $f : A \to B$.

E.1 Postnikov section. This is perhaps the oldest homotopically idempotent functor although the usual construction does not present it as a continuous or simplicial functor. The above construction presents a simplicial (continuous) version of it. One takes the map to be $u_{n+1} : * \to S^{n+1}$. Then $P_n X = P_{S^{n+1}} = \mathbf{L}_{u_{n+1}}$ since $\mathrm{map}_*(S^{n+1}, X) \simeq \Omega^{n+1} X$ and $\Omega^{n+1} X \simeq *$ is equivalent to $\pi_i X \simeq 0$ for $i \geq n+1$. Many of the properties of the general nullification functor with respect to suspensions are generalizations of analogous properties of this case.

E.2 Sullivan–Quillen localizations: This functor inverts in a natural way a set of primes P by tensoring $\pi_* X$ with $\mathbb{Z}[P^{-1}]$ for all $n \geq 2$ and X 1-connected. To get this effect we can choose the map f as the wedge $f : \bigvee_{p \in P} S^2 \xrightarrow{\vee p} \bigvee_{p \in P} S^2$, of all the maps of degree $p : S^2 \to S^2$. Notice, however, that $\mathbf{L}_P$ is now defined for an arbitrary space X and is equivalent to the 'fibrewise Sullivan localization' of the fibration $\tilde{X} \to X \to K(\pi_1 X, 1)$.

E.3 Bousfield–Kan localizations and completions: These localizations agree with Sullivan's for 1-connected spaces of finite type and subrings of the rationals. Although they are always well defined and simplicial, these functors are not localizations since , in general, the Bousfield–Kan functor R_∞ is not idempotent. In any subcategory on which R_∞ is idempotent it agrees with $\mathbf{L}_f$ for a well chosen f. Thus if one takes the wedge of degree p maps over a set P of primes: $f : \bigvee_P S^1 \xrightarrow{\vee p} \bigvee_P S^1$, this will render all the homotopy groups uniquely p-divisible for all $p \in P$. It will also render them divisible by certain elements in the group ring of the fundamental group as was shown by Peshke, see [Ca-P]. This is closely related to the work of Baumslag [Baum-1]. For nilpotent space X the localization $\mathbf{L}_f X$ with respect to this map of circles is still nilpotent and unique divisibility implies $\mathbf{L}_p = \mathbf{L}_f$-local. So one gets a new localization when localizing non-simply connected spaces with respect to the degree p map on the circle S^1 . One gets a space for which the degree p map on the loop space is a homotopy equivalence (disregarding the loop structure). In terms of homotopy groups this is a stronger condition than unique divisibility. It is equivalent to the condition that all elements in the group ring of the fundamental group of the form $1 + \xi + \xi^2 + \cdots + \xi^p$ act on the higher homotopy groups by automorphism.

For other rings such as $\mathbb{Z}/p\mathbb{Z}$, one can use other maps (see [DF-4]). The main advantage of R_∞ over $\mathbf{L}_f$ is the existence of a manageable (Adams-like) spectral sequence converging to $\pi_i \operatorname{map}(W, R_\infty X)$.

E.4 Bousfield homological localization: Let h_* be any generalized homology. It follows from the discussion [B-1] [DF-4] that if a space X is f-local for any map $f : A_1 \to A_2$ that induces an h_*-isomorphism between two spaces of cardinality not bigger than the cardinality of $h_*(S^0)$, then X is h_*-local i.e. it is local with respect to <u>any</u> map that induces an isomorphism on h_*. Therefore, in order to invert all h_*-equivalence we choose $f : A \to B$ to be the wedge over 'all' h_*-isomorphisms between small enough spaces—taking one copy of each homotopy type, of course. The resulting functor will be homotopy equivalent to Bousfield's $\mathbf{L}_{h_*}$. This applies in particular to integral homology. The advantage of the present construction is that $\mathbf{L}_f$ here is simplicial (compare [B-3]). Also, we will see that our ability to consider $\mathbf{L}_{\Sigma f}$, where $\Sigma f : \Sigma A \to \Sigma B$ is the usual suspension of the above 'monster' map, is very useful in proving fibre lemmas for $\mathbf{L}_{h_*}$.

E.5 Quillen's plus construction: This functor $X \to X^+$ can be expressed as $\mathbf{P}_A$, where A is a large acyclic space with respect to integral homology. Namely we take A to be the wedge of all acyclic spaces whose cardinality (i.e. of the set of cells) is the same as $H_0(*, \mathbb{Z})$. Since any $H_*(-, \mathbb{Z})$-acyclic space is a direct limit of its countable acyclic subcomplexes, it is not hard to see, using the commutation rules for homotopy colimits, that $\mathbf{P}_A X \simeq X^+$.

In fact, one can use this interpretation to define a version of the plus construction for any homology theory E_* simply by taking $X_E{}^+$ to be $\mathbf{P}_{A(E)} X$, where $A(E)$ is a large enough wedge of E_*-acyclic spaces. Theorem 1.H.2 below says that the homotopy fibre of the map $X \to X_E{}^+$ is always E_*-acyclic.

E.6 Periodicity functors: Mahowald, Ravenel and others consider algebraic and geometrical v_1-periodic families and spaces. Let $M^n(\mathbb{Z}/p\mathbb{Z})$ denote the Moore space with a top cell at dimension n. Homotopically we can consider the mod-p homotopy group $[M^n(\mathbb{Z}/p\mathbb{Z}), X] \equiv \pi_n(X, \mathbb{Z}/p\mathbb{Z})$ and the v_1-periodicity operator induced by the Adam's map $v_1 : M^{n+q}(\mathbb{Z}/p\mathbb{Z}) \to M^n(\mathbb{Z}/p\mathbb{Z})$ ($q = 2p-2$ for an odd prime p,[C-N]).

A v_1-periodic space, naively speaking, is a space for which v_1 induces an isomorphism on $\mathbb{Z}/p\mathbb{Z}$-homotopy. The functor $\mathbf{L}_{v_1}$, localization with respect to v_1, turns every space into a v_1-periodic one. In Chapter 8 below we give an approach to this functor largely following Bousfield and Thompson.

E.7 FURTHER EXAMPLES: Notice that if we localize with respect to any null-homotopic map $u = A \to B$, we get an $(A \vee B)$-null space. If we localize with respect to a nilpotent self-map $\Sigma^k W \to W$, we get the same result as localization with respect to the null map, thus we get the functor P_W. We will later see that properties of P_W are closely related to the Hopkins–Smith theory of self-maps $\Sigma^k W \to W$ for finite p-torsion spaces [Ho].

F. Fibrewise localization

From a general point of view any functor $\mathbf{F} : \mathcal{S} \to \mathcal{S}$ that preserves weak homotopy equivalences can be applied fibrewise to fibre maps $p : E \to X$ by 'applying $\mathbf{F}$ to each fibre'. It is easier to construct a version fibrewise application of a functor that is natural only up to homotopy compare [B-4] and sometimes that is the only way we present here (3.E.2). This means that the construction is functorial in the homotopy category. In our present applications that will suffice. Still, it is nice to have a functorial, 'rigid' fibrewise localization that will have universal properties in the category of spaces over B similar to those of $\mathbf{L}_f$ in the category of spaces. This can be done both for simplicial sets and in the topological category.

It takes a bit of pushing and pulling to define the fibrewise application of a functor functorially in the category of simplicial sets, and this is done below. By further homotopy pushouts and pullbacks along weak equivalences using the singular and realization functors, we show below how to induce a functorial fibrewise localization from simplicial sets to topological spaces (F.6), thus constructing these localizations in the latter category. In (F.7) we give a rough outline of another direct functorial fibrewise localization in a topological setting without using (strict) continuity. Continuity 'up to homotopy' is always there for homotopy functors. An extended and careful treatment of these issues is given in [HH].

So first we formulate *in a special case* an existence theorem for fibrewise application, up to homotopy, of simplicial functors.

F.1 THEOREM (FIBREWISE APPLICATION OF COAUGMENTED HOMOTOPY FUNCTORS): *Let* Id $\longrightarrow \mathbf{F} : \mathcal{S} \to \mathcal{S}$ *be a simplicial coaugmented functor. Every fibration* $X \to E \to B$ *of connected spaces can be mapped via a homotopy commutative ladder (F.1.3) to a fibration over* B *of the form:* $\mathbf{F}X \to \overline{E} \to B$. *If* $X \longrightarrow \mathbf{F}X$ *is an* f*-equivalence then in this ladder the map* $E \longrightarrow \overline{E}$ *is an* f*-equivalence.*

REMARK: In fact it is enough to assume that our functor **F** is a homotopy functor, i.e. that it preserves weak equivalences. In that case the theory of Dwyer and Kan [D-K-1] shows that it is essentially a simplicial functor.

Proof: Notice that once one constructs the ladder the last statement in (F.1) follows at once from example (1.D.8). We use the simplicial structure of **F**: This implies that for any space there is a map $\mathrm{aut}X \to \mathrm{aut}\mathbf{F}X$ from the space of homotopy self-equivalence of X to that of **F**X. This map is the restriction of the corresponding map on the spaces of all self-maps to the components of self-equivalences.

Now for a connected X there is always a universal fibration $X \longrightarrow \mathrm{Baut}^{\bullet}X \longrightarrow \mathrm{Baut}X$ where $\mathrm{aut}^{\bullet}X$ is the monoid of pointed self-equivalences of X with respect to any choice of base point $* \in X$. That fibration comes from the evaluation map and every fibration with fibre X is, up to fibre homotopy equivalence, a pullback from this fibration [DF-Z-1].

Now notice that for any fibration $X \to E \to B$ of connected spaces we have a commutative diagram in which the natural map from E to the homotopy pullback is a homotopy equivalence. Thus, up to homotopy, the following is a homotopy pullback diagram:

(F.1.1)
$$\begin{array}{ccc} E & \longrightarrow & \mathrm{Baut}^{\bullet}X \\ \downarrow & & \downarrow \\ B & \longrightarrow & \mathrm{Baut}X \end{array}$$

This presents our fibration up to homotopy as a pullback from a canonical fibration that depends only on the choice of base point $* \in X$. Notice that the map out of B is defined by the given fibration only up to homotopy. In order to complete the construction we need also a canonical map $\mathrm{Baut}^{\bullet}X \longrightarrow \mathrm{Baut}^{\bullet}\mathbf{F}X$. Choose a base point $* \in \mathbf{F}X$ by composition with the coaugmentation. This guarantees that the map of spaces of self-equivalences restricts to a corresponding map of pointed self-equivalences. Thus there is a commutative diagram:

$$
\begin{array}{ccc}
X & \dashrightarrow & \mathbf{F}X \\
\downarrow & & \downarrow \\
\mathrm{Baut}^{\bullet}X & \longrightarrow & \mathrm{Baut}^{\bullet}\mathbf{F}X \\
\downarrow & & \downarrow \\
\mathrm{Baut}X & \longrightarrow & \mathrm{Baut}\mathbf{F}X
\end{array}
\tag{F.1.2}
$$

Once we have the above (F.1.1) pullback presentation, we can compose the bottom map with the map $\mathrm{Baut}X \to \mathrm{Baut}\mathbf{F}X$ obtained by taking the classifying space functor on the map above. We now can define $\overline{E}$ as the pullback from the universal fibration with fibre $\mathbf{F}X$ using (F.1.2). This gives us a homotopy commutative diagram claimed by the theorem:

$$
\begin{array}{ccccc}
X & \longrightarrow & E & \longrightarrow & B \\
\downarrow & & \downarrow & & \downarrow \\
\mathbf{F}X & \longrightarrow & \overline{E} & \longrightarrow & B
\end{array}
\tag{F.1.3}
$$

The following proposition is now clear from the construction.

F.1.4 Corollary: *If the fibration $X \to E \to B$ is classified by the map $B \to \mathrm{Baut}X$, then $\mathbf{F}X \to \overline{E} \to B$, where $\overline{E}$ is the fibrewise application of $\mathbf{F}$ to E over B, is classified by $B \to$ Baut$X \to$ Baut$\mathbf{F}X$, where the second map is induced by the simplicial structure of the functor $\mathbf{F}$.*

F.2 Rigid and functorial fibrewise localization, universal properties.

We will now use a slightly more sophisticated technique, but still a straightforward one, to construct a fibrewise localization in a functorial way by literally applying our functor to the inverse image in E of each simplex in B. This technique is useful also for other purposes as will be seen in Chapter 9. For simplicity, one may assume that B is connected and that is all we need below. But this is by no means a necessary restriction. In fact this can be done with every functor $\mathbf{F}$ that preserves weak equivalences.

This can also be done in the framework of 'simplicial localizations' [D-K-1] that guarantees that 'up to equivalence' there is always a map $\mathrm{Hom}(X, Y) \to$

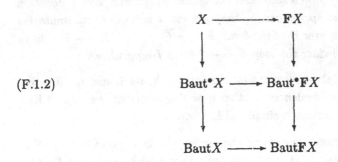

Hom($\mathbf{F}X, \mathbf{F}Y$) since one can recover the homotopy type of Hom(X, Y) from the collection of weak equivalences. In our more elementary framework of simplicial sets this can be done more concretely by considering the total space E up to homotopy equivalence as a homotopy colimit (unpointed) of a diagram of the fibres over the indexing category that can be taken as the category of simplices in B with morphisms being simplicial face and degeneracy maps.

The basic idea is simple and appeared among other places in works about homotopy theory of small categories [L-T-W]: One first decomposes the base space B into a diagram of its simplices. This way B appears as a colimit of a diagram ΓB whose shape (i.e. the underlying small category) is determined by B itself and where the spaces in this diagram are the simplices $\Delta[n]$ for $n = 0, 1, 2, \ldots$ which together give B as a space (simplicial set or complex). Now one can consider the inverse images of the simplices of B in E. The collection of all these inverse images forms a diagram over ΓB consisting of spaces that are homotopy equivalent to the fibres over the corresponding component; in this diagram all maps are weak homotopy equivalences. Now since our functor $\mathbf{F}$ preserves every weak equivalence, after applying the functor to the diagram of inverse images we still get a diagram of weak equivalences. Taking the homotopy colimit of this diagram we obtain the desired new total space up to homotopy equivalence. The only problem is that, as it stands, it maps naturally to the classifying space of ΓB, a space that is homotopy equivalent to B, and it does not map to B itself. Now we use some pushing and pulling to bring it, in a natural way, over B. We will need a few general lemmas about decomposing fibrations. This information is essentially contained in Quillen's Theorem B [Q-2] or in V. Puppe's theorem [Pu] about homotopy colimits of fibrations; see Appendix HL at the end.

All this is a straightforward procedure. It is essential, though, for this approach that our functor will be defined in the unpointed category of spaces.

F.3 THEOREM: *Let $\mathbf{F} : \mathcal{S} \to \mathcal{S}$ be any coaugmented functor that sends contractible spaces to contractible spaces and preserves weak equivalences. Let $g : E \to B$ be a fibre map. There is a natural commutative diagram over B:*

$$\begin{array}{ccc} X_0 & \longrightarrow & \mathbf{F}X_0 \\ \downarrow & & \downarrow \\ E & \longrightarrow & \overline{E} \\ \downarrow & & \downarrow \\ B & = & B \end{array}$$

where X_0 is a homotopy fibre over any component of B and $\mathbf{F}X_0$ the corresponding homotopy fibre.

Example: The above construction has been used fruitfully for functors other than localization. For example, Dwyer in [Dw-1] applies such a theorem for the case where $\mathbf{F}$ is the function complex with the domain being any space W, $\mathbf{F}_W = \mathbf{F}X = \mathrm{map}(W, X)$, that is coaugmented by the constant maps. Thus for every fibration $X \to E \to B$ we get an associated fibration $\mathrm{map}(W, X) \to \overline{E} \to B$, and a corresponding commutative diagram.

Proof: We use the discussion in Appendix HL regarding the decomposition of fibre maps into free diagrams. Given the map $E \to B$ we get a map of diagrams over the indexing diagram $I = \Gamma(B)$, namely $\underset{\sim}{E} \to \underset{\sim}{B}$, where $\Gamma(B)$ is the small category of all the simplices of B with maps being the simplicial maps. We also have a natural diagram of weak equivalences, where I is a shorthand here for the diagram $\Gamma(B)$, i.e. $I = \Gamma(B)$:

$$\text{(F.3.1)} \qquad \begin{array}{ccc} \operatorname{hocolim}_I \underset{\sim}{E} & \xrightarrow{\simeq} & \operatorname{colim} \underset{\sim}{E} = E \\ \downarrow & & g \downarrow \\ \operatorname{hocolim}_I \underset{\sim}{B} & \xrightarrow{\simeq} & \operatorname{colim} \underset{\sim}{B} = B \end{array}$$

We now apply the functor $\mathbf{F}$ object-wise to both diagrams $\underset{\sim}{E}$ and $\underset{\sim}{B}$ to get the diagrams maps $\underset{\sim}{E} \to \mathbf{F}\underset{\sim}{E}$ and $\underset{\sim}{B} \to \mathbf{F}\underset{\sim}{B}$. Taking homotopy colimits we obtain a natural commutative square (with $I = \Gamma(B)$):

$$\text{(F.3.2)} \qquad \begin{array}{ccc} \operatorname{hocolim}_I \underset{\sim}{E} & \longrightarrow & \operatorname{hocolim}_I \mathbf{F}\underset{\sim}{E} \\ \downarrow & & \downarrow g_{\mathbf{F}} \\ \operatorname{hocolim}_I \underset{\sim}{B} & \xrightarrow{b} & \operatorname{hocolim}_I \mathbf{F}\underset{\sim}{B} \end{array}$$

Now since $\mathbf{F}(pt)$ is contractible and $\mathbf{F}$ respects weak equivalences, the diagram $\mathbf{F}\underset{\sim}{B}$ is a diagram of contractible spaces.

In other words, the map of diagrams over $\Gamma(B)$, namely $\underset{\sim}{B} : \underset{\sim}{B} \to \mathbf{F}\underset{\sim}{B}$, induces a weak equivalent on each object $\sigma \in \Gamma(B)$, therefore upon taking homotopy colimit we get a weak equivalence, which is the map b. Now let $\overline{E}_1$ be the homotopy pullback in (F.3.2) of $\operatorname{hocolim}_I \mathbf{F}\underset{\sim}{E}$ along b. We have a fibre map $g_1 : \overline{E}_1 \to \operatorname{hocolim}_I \underset{\sim}{B}$ whose homotopy fibres over each component are the same as those of $g_{\mathbf{F}}$ in (F.3.2) above. We now combine the above, composing g_1 with the bottom map in (F.3.1) to get a diagram of spaces over B:

$$(\text{F.3.3})\qquad \begin{array}{ccccc} E & \xleftarrow{\cong} & \operatorname{hocolim}_{I} \underset{\sim}{E} & \xrightarrow{\ell} & \overline{E}_1 \\ \downarrow & & \downarrow & & \downarrow \\ B & = & B & = & B \end{array}$$

where the map ℓ is obtained by the strict commutativity of (F.3.2) as usual, see Appendix HL. The weak equivalence on the top left in (F.3.3) is the same one as in (F.3.1). We now finally define $\overline{E}$ to be the homotopy pushout of the top arrows in (F.3.3). Since there is a pushout along a weak equivalence, the natural map $\overline{E}_1 \to \overline{E}$ is a weak equivalence and its homotopy fibres over B are weakly equivalent to those of $\overline{E}_1 \to B$. Since these are equivalent to $\mathbf{F}(g^{-1}(\sigma))$ for any $\sigma \in B$, we get the result claimed. Since all the steps were natural, the resulting maps $E \to \overline{E}$ and $\overline{E} \to B$ are functorial in $g : E \to B$. This completes the proof.

In fact the argument in the proof of (F.3) proves a bit more:

F.4 Theorem: *Let $E \to B$ be a fibre map and let $a : E \to \overline{E}$ be the fibrewise localization obtained by applying $\mathbf{L}_f$ fibrewise. Then a is an f-local equivalence.*

Proof: Apply example (1.D.7). Alternatively, directly by construction, the map $E \to \overline{E}$ is homotopy equivalent to the map $\operatorname{hocolim}_I \underset{\sim}{E} \to \operatorname{hocolim}_I \mathbf{L}_f \underset{\sim}{E}$, with $I = \Gamma(B)$. But by Theorem D.2 above this is an f-local equivalence, since for each object $\sigma \in \Gamma(B)$ we get a localization map which is always an f-local equivalence (C.5.1) (D.2.1).

F.5 Universal properties: Fibrewise localization can be easily seen to have the appropriate universal properties: The map $E \to \overline{E}$ as a map over the space over X is universal among all maps of E to spaces over X that have an f-local space as a (homotopy) fibre.

F.6 Fibrewise localization in the topological category: In order to pull back the above construction and in fact other constructions such as the localization in the absolute case, from simplicial sets (denoted $\mathcal{SS}$) to topological spaces (Top) over a fixed space B, we use the pair of adjoint functors $|-|$=realization: $\mathcal{SS} \longrightarrow$ Top and Sing, the singular functor, going in the other direction. There is a natural transformation which is a weak equivalence $u : |\text{Sing}| \to \text{Id}$ and we use it to construct the localization in Top using the fibrewise localization in $\mathcal{SS}$. Let $E \to B$ be a fibre map in Top. Then $\text{Sing}(E) \to \text{Sing}(B)$ is a fibre map in $\mathcal{SS}$. We apply to it fibrewise localization in the category of simplicial sets and get $\overline{\text{Sing}(E)} \to \text{Sing}(B)$. One then can take the realization of the localization map over $\text{Sing}(B)$ and compose it into B by the above transformation to the identity u. So one obtains a pushout diagram where the top left arrow is a cofibre map and the right one is a weak equivalence:

$$\text{(F.6.1)}\qquad \begin{array}{ccccc} \overline{|\mathrm{Sing}(E)|} & \xleftarrow{\text{cof}} & |\mathrm{Sing}(E)| & \xrightarrow{\simeq} & E \\ \downarrow & & \downarrow & & \downarrow \\ |\mathrm{Sing}(B)| & \xleftarrow{=} & |\mathrm{Sing}(B)| & \xrightarrow{\simeq} & B \end{array}$$

Now the pushout, denoted say by $\overline{E}$ of the top two arrows in (F.6.1) (along the cofibration cof), is the desired space over B. This space is of course weakly equivalent to the realization of the fibrewise localization in the simplicial category $\mathcal{SS}$.

F.7 Another approach to fibrewise localization without continuity

Given $\mathbf{L}_f : \mathcal{S} \to \mathcal{S}$ associated to f: $V \longrightarrow W$, we can extend it to fibrewise localization by the following method without assuming a simplicial structure on the given functor $\mathbf{L}_f$. It was shown in [HH] for pointed spaces this approach can work only for a fibration with connected fibre since it localizes all the components of the fibre.

We start by choosing a base point in the base space B of the given fibration sequence $F \longrightarrow E \longrightarrow B$. One applies $\mathbf{L}_f$ only to the fibre over the base point, then one turns, naturally, the resulting map into a fibre map. Again apply $\mathbf{L}_f$ to the fibre over the base point $(*)$, etc., doing this λ times for $\lambda = \lambda(V \amalg W)$ to give fibrewise localization. Here is an outline of an argument: Let $\mathbf{L}_f$ be a given localization functor in the category of spaces. Given a fibre map $E \to B$ one first applies $\mathbf{L}_f$ to the inverse image of the base point : Define E' to be the pushout along cofibration of $\mathbf{L}_f F \longleftarrow F \longrightarrow E$. Thus E' maps naturally to B see [HL]. Now define E_1 to be the functorial space that turns the map $E' \to B$ into a fibre map so that $E_1 \to B$ is a fibre map. Now the fibre of this map over the base point $* \in B$ is no longer f-local so we repeat the last two steps again, and by a transfinite induction we construct a long telescope. We will need to carry on until a well chosen transfinite ordinal as in (B.3) above. Now the transfinite homotopy colimit F_λ is a long direct limit of f-local spaces, since every other place in the telescope, namely the spaces $F_\beta{}'$, is f- local and taking fibres commute with this colimit (HL), and for this ordinal taking mapping spaces commute with direct limits too,see (B.3) above. Furthermore, all the maps in the telescope of inverse images of the base point in B are f-equivalences. This follows directly from (D.3) above and the expression of the homotopy fibre as a homotopy colimit of the inverse images of simplices in the base (9.B.1 below): Every other map in this tower is localization so certainly an f-equivalence and for the other family one uses Proposition F.7.1 below:

F.7.1 PROPOSITION: *Let $X \to Y$ be any map over a pointed space B:*

$$\begin{array}{ccc} X & \xrightarrow{g} & Y \\ {\scriptstyle p}\downarrow & & \downarrow{\scriptstyle q} \\ B & \xrightarrow{=} & B. \end{array}$$

If for any $\sigma \in B$ the map g induces an f-equivalence $\mathbf{L}_f(p^{-1}(\sigma)) \simeq \mathbf{L}_f(q^{-1}(\sigma))$ then g is an f-equivalence, moreover the map g induces an f-equivalence on the homotopy fibres $\mathbf{L}_f\mathrm{Fib}(p) \simeq \mathbf{L}_f\mathrm{Fib}(q)$.

Proof: This is clear from the fact that hocolim preserves f-equivalence see Section D above and from the construction of the homotopy fibre and of the total space as an (unpointed) homotopy colimit of the diagram of the inverse images of points or simplices in the base indexed by the small category ΓB associated with B see (F.2) and Appendix HL. ∎

G. Proof of elementary facts

We give short proofs of the properties listed in section A.8 above.

e.1 If any f-local space is g-local and vice versa, then $\mathbf{L}_f$ and $\mathbf{L}_g$ satisfy the same universality condition. If $\mathbf{L}_f \cong \mathbf{L}_f$, then an f-local space X satisfies $X \cong \mathbf{L}_f X \cong \mathbf{L}_g X$ so it is also g-local.

e.2 This can be seen by adjunction: $\mathrm{map}(A, \mathrm{map}(X,T)) \simeq \mathrm{map}(X, \mathrm{map}(A,T))$. Since by assumption T is f-local one can replace in the latter A by B and get the desired (unpointed) result. For the pointed case one argues similarly or use for any choice of base point in T the fibration:

$$\mathrm{map}_*(X,T) \to \mathrm{map}(X,T) \to T.$$

Since we can assume without loss of generality that T is connected, the statement follows by (e.3).

e.3 Let $\underset{\sim}{X} : I \to S$ be a diagram of f-local spaces. We must check that the map induced by $f : A \hookrightarrow B$ on $\mathrm{holim}_I \underset{\sim}{X}$ namely

$$\mathrm{map}(f, \mathrm{holim}_I \underset{\sim}{X}) : \mathrm{map}(B, \mathrm{holim}_I \underset{\sim}{X}) \to \mathrm{map}(A, \mathrm{holim}_I \underset{\sim}{X}),$$

is a weak equivalence. But this map is $\mathrm{holim}_I \mathrm{map}(f, \underset{\sim}{X})$ by the basic property of homotopy limits. Now by assumption this map of the diagram induces a weak equivalence on $\underset{\sim}{X}(i)$ for each $i \in I$. By the fundamental property of homotopy limits [B-K], Appendix HL at the end, we get the desired weak equivalence.

e.4 We construct maps as follows, for which, by universality (divisibility), one gets immediately that they are homotopy inverse to each other:

$$\ell : \mathbf{L}_f(X \times Y) \leftrightarrows \mathbf{L}_f X \times \mathbf{L}_f Y : r.$$

To get ℓ we extend by universality the product of the two coaugmentations $X \times Y \to \mathbf{L}_f X \times \mathbf{L}_f Y$, since by (e.3) the product of f-local spaces is f-local. To get r we start with the coaugmentation $X \times Y \to \mathbf{L}_f(X \times Y)$. Adjunction gives a map $X \to \mathrm{map}(Y, \mathbf{L}_f(X \times Y))$. Since the range is f-local by (e.2), this map factors through $\mathbf{L}_f X$. By adjunction again we get a map $Y \to \mathrm{map}(\mathbf{L}_f X, \mathbf{L}_f(X \times Y))$, which again factors through $\mathbf{L}_f Y$, yielding by another adjunction the desired map.

e.5 For any map $f : A \to B$ of pointed spaces, $f \wedge W : A \wedge W \to B \wedge W$ satisfies $\mathrm{map}_*(f \wedge W, T) \cong \mathrm{map}_*(f, \mathrm{map}_*(W, T))$. Now use (e.1), (e.2) to get the conclusion.

e.7 For $W = S^1$ this follows directly form adjointness of suspension of loop functors. For a general CW complex one argues by induction on skeleta. In order show that $\mathrm{map}_*(W, X)$ is f-local, one uses induction on the skeleton of W. Since W is n-connected, only $(n+i)$-dimensional spheres for $i \geq 1$ are used in the inductive construction $S^{n+i} \to W \to W'$. Now use the usual theorem about mapping of a cofibration into a space (Appendix HL below) to get $\mathrm{map}_*(W', X)$ as the homotopy fibre of maps between two f-local spaces. The infinite union is treated similarly, since the needed function complex appears as the homotopy (inverse) limit of f-local spaces (by Appendix HL), so by (e.3) it is also f-local.

e.8 This follows directly from (e.2) and (e.7).

e.9 This follows directly from universality .

e.10 This follows from (e.7) for $W = S^1$, which is the first case. In general, in order to show an equation $\mathbf{L}_f X \simeq *$, it is sufficient by universality to show that for any f-local space T the pointed function space $\mathrm{map}_*(X, T)$ is contractible. Let T then be a $\Sigma^{n+1} f$-local space. Then $\mathrm{map}_*(X \wedge W, T) \simeq \mathrm{map}_*(X, \mathrm{map}_*(W, T)) \simeq *$ since, by the connectivity of W, it follows that $\mathrm{map}_*(W, T)$ is f-local (e.7).

e.11 This is immediate from (e.2): Simply use T a low dimensional Eilenberg–Mac Lane space $K(G, i)$ for $i \leq n$ any any G. See (A.1.1). Such a space is f-local and so for any G and above i we have $* \simeq \mathrm{map}(X, K(G, i)) \simeq \mathrm{map}(\mathbf{L}_f X, K(G, i))$. Hence the claimed connectivity.

H. The fibre of the localization map

In this section we apply the notion of fibrewise localization in order to prove two central properties of f-localizations:

H.1 THEOREM: *If $F \to E \xrightarrow{p} X$ is a fibration and $\mathbf{L}_f F \simeq *$, then $\mathbf{L}_f(p) : \mathbf{L}_f E \to \mathbf{L}_f X$ is a homotopy equivalence.*

H.2 THEOREM: *Let $\mathbf{P}_W = \mathbf{L}_{W \to *}$ be the W-nullification functor with respect to any space W. Let X be a pointed connected space. Then $\mathbf{P}_W \overline{\mathbf{P}}_W X \simeq *$, where $\overline{\mathbf{P}}_W X$ is the homotopy fibre of the nullification $X \to \mathbf{P}_W X$.*

H.3 THEOREM: *Let X be a pointed connected space. Then $\mathbf{L}_f \overline{\mathbf{L}}_{\Sigma f} X \simeq *$ where $\overline{\mathbf{L}}_g X$ denotes the homotopy fibre of the localization map $X \to \mathbf{L}_g X$.*

Remark: Notice that (H.3) gives a 'weaker statement' than (H.2).

Proof of H.1: If one fibrewise localize E over X, one gets a map $\overline{E} \to X$ whose fibre $\mathbf{L}_f F \simeq *$ is contractible, thus $\mathbf{L}_f \overline{E} \simeq X$. But $\mathbf{L}_f \overline{E} \simeq \mathbf{L}_f E$ by (F.4), so $\mathbf{L}_f E \simeq \mathbf{L}_f X$ as needed.

Proof of H.2: We may assume that the map $X \to \mathbf{P}_W X$ is a fibre map. Consider the following fibrewise localization diagram:

(H.2.1)
$$\begin{array}{ccccc}
F & \longrightarrow & X & \xrightarrow{\ \ell\ } & \mathbf{P}_W X \\
\downarrow & & \downarrow{\scriptstyle \tilde{\sigma}} & \swarrow{\scriptstyle \sigma} & \downarrow{\scriptstyle =} \\
\mathbf{P}_W F & \longrightarrow & \overline{X} & \xrightarrow[\ \tilde{\ell}\ ]{} & \mathbf{P}_W X
\end{array}$$

By (A.3) above $\overline{X}$ is in fact W-null and by universality one has a map σ with $\tilde{\sigma} \sim \sigma \circ l$. Since $\tilde{\ell} \circ \sigma \circ \ell \sim \tilde{\ell} \circ \tilde{\sigma} \sim \mathrm{id} \circ \ell$ we have by uniqueness of factorization (C.3) $\tilde{\ell} \circ \sigma \sim \mathrm{id}$. Therefore σ is a section of the fibre map $\tilde{\ell}$. This means, by the usual long exact sequence of fibration, that $\mathbf{P}_W F \to \overline{X}$ induces a one-to-one map on pointed homotopy classes $[V, \mathbf{P}_W F]_* \to [V, \overline{X}]_*$ for any space V and, in particular, for $V = F$. Since $F \to \overline{X}$ factors through the base space $\mathbf{P}_W X$ of the fibration, it must be null homotopic and thus $F \to \mathbf{P}_W F$ is also null homotopic. Idempotency of $\mathbf{P}_W$ (C.4) now gives $\mathbf{P}_W F \simeq *$, as claimed.

Proof of Theorem H.3: The proof is a slight variation on that of (H.2) above: Consider the fibration sequence in which all the maps are the natural ones:

(H.3.1)
$$\begin{array}{ccccc} F & \longrightarrow & X & \xrightarrow{\ell} & \mathbf{L}_{\Sigma f}X \\ \downarrow & & \downarrow b & \swarrow \sigma & \downarrow \\ \mathbf{L}_f F & \longrightarrow & \overline{X} & \xrightarrow[\bar{\ell}]{} & \mathbf{L}_{\Sigma f}X \end{array}$$

By elementary fact e.7 in (A.8) above the total space $\overline{X}$ is Σf-local.

By the universality property of the map $X \to \mathbf{L}_{\Sigma f}X$, the map $X \to \overline{X}$ factors through it up to homotopy. This gives a map σ. But this means that the fibration $\overline{X} \to \mathbf{L}_{\Sigma f}X$ has a cross-section. Notice that by uniqueness of factorization we shall get $\bar{l} \circ \sigma \sim \mathrm{id}$ from $\bar{l} \circ \sigma \circ l \sim l$, but $\sigma \circ l \sim b$ and $\bar{l}b \sim l$ by construction (compare proof of H.2). Thus the map $\mathbf{L}_f F \to \overline{X}$ induces a one-to-one map on pointed homotopy class $[W, \mathbf{L}_f F] \to [W, \overline{X}]$ for any space W, in particular for $W \sim F$. Notice, however, that the map $F \to \overline{X}$ factors through the base of the fibration $\mathbf{L}_{\Sigma f}X$, and therefore it is a null homotopic map. But this means that $F \to \mathbf{L}_f F$ is null homotopic, and therefore the idempotency of $\mathbf{L}_f$ gives that $\mathbf{L}_f F \sim *$, as claimed.

H.4 Remark. It is essential in (H.3) that $\mathbf{L}_f F \simeq *$ rather than $\mathbf{L}_{\Sigma f} F \simeq *$ holds. The latter is, in general, not true as can be observed by taking f to be the degree p-map between two 2-spheres $\Sigma f = p : S^2 \to S^2$. If $X = S^2$ itself $\pi_1 F$ will be $\mathbb{Z}[\frac{1}{p}]/\mathbb{Z}$, a p-torsion group. But it is clear from the construction of $\mathbf{L}_{\Sigma f}$ that $\pi_1 \mathbf{L}_{\Sigma f} F = \pi_1 F \neq 0$. However, if we had taken $f = p : S^1 \to S^1$, then with the same F we would have $\mathbf{L}_f F \simeq *$.

H.5 Remark. An inductive application of (H.1) gives for $W = K(\mathbb{Z}/p\mathbb{Z}, n)$ and any positive natural numbers n and k:

$$\mathbf{P}_W K(\mathbb{Z}/p^k\mathbb{Z}, n) \simeq (*).$$

2. AUGMENTED HOMOTOPY IDEMPOTENT FUNCTORS

A. Introduction, A-equivalence

In this chapter we treat certain augmented functors **F**, namely functors equipped for each space X with a natural map $\mathbf{F}X \longrightarrow X$. In fact these turn out to be kind of colocalization functors — they are closely related to homotopy fibres of localization maps. Just as one can introduce localizations by first defining f-local spaces, and f-local equivalences, we start here with a class of maps with respect to which the new functors are universal.

It is interesting to notice that assuming idempotency these augmented functors, as opposed to coaugmented ones, are non-trivial only on *pointed spaces* as will shortly be seen (A.3.4). So in the present chapter we work only in the category $\mathcal{S}_*$ of pointed spaces. Typical examples of these functors are universal and n-connected covers for $n \geq 1$.

The first definition attempts to capture the concept of 'the information in a space X that can be detected by maps from a fixed space A.'

A.1 DEFINITION: *A pointed map g: $W \to X$ of fibrant spaces is called an A-homotopy equivalence or simply A-equivalence, where $A \in \mathcal{S}_*$ is cofibrant, if it induces a (weak) homotopy equivalence on the pointed function complex*

$$\text{map}_*(A, g)\colon \text{map}_*(A, W) \to \text{map}_*(A, X).$$

A.1.1 REMARK: In case the spaces involved in g are not fibrant, we ask for a weak equivalence on the function complexes of the associated realization or associated fibrant objects.

It is pointless to consider the unpointed version of an A-equivalence since it would be, in general, a weak homotopy equivalence. Any space in $\mathcal{S}$ is a retract of $\text{map}(A, S)$ if $A \neq \emptyset$ and a retract of an equivalence is an equivalence.

A.2 DEFINITION: *We say that a functor T: $\mathcal{S}_* \to \mathcal{S}_*$ is homotopy idempotent and augmented if it comes with a natural augmentation $a = a_X : TX \to X$ and the two maps $T(a), a_T : T^2X \to TX$ are homotopic to each other and both are homotopy equivalences.*

A.3 EXAMPLES. 1. A well known example of an augmented functor is the universal cover functor $\tilde{X} \to X$. More generally, we have the n-connected functor cover $X\langle n\rangle \to X$ for $X \in \mathcal{S}_*$. Notice that the latter is an S^{n+1} equivalence in the sense of (A.1).

2. Let $A \in \mathcal{S}_*$ and let $\overline{\mathbf{P}}_A X \to X$ be the fibre of the coaugmentation $X \to \mathbf{P}_A X$ from (1.H.2) above. We claim that this is a homotopy idempotent augmented functor $\overline{\mathbf{P}}_A : \mathcal{S}_* \to \mathcal{S}_*$. Idempotency is an immediate consequence of (1.H.2) above, since $\mathbf{P}_A \overline{\mathbf{P}}_A \simeq *$ means $\overline{\mathbf{P}}_A \overline{\mathbf{P}}_A \simeq \overline{\mathbf{P}}_A X$. Now it follows directly from this that the map $\overline{\mathbf{P}}_A X \to X$ is universal among maps $W \to X$ with $\mathbf{P}_A W \simeq *$, i.e. any such map factors up to homotopy through $\overline{\mathbf{P}}_A X$ uniquely up to homotopy. The *acyclic functor* i.e. the homotopy fibre of Quillen's plus construction belongs to this class of examples [DF-6]. We are not aware of a similar characterization of the fibre $\overline{\mathbf{L}}_f X$ of $X \to \mathbf{L}_f X$. Certainly $\overline{L}_f$ is not homotopy idempotent for $f\colon S^n \to S^n$ $(n \geq 1)$, for example (1.H.4).

3. Let Top_* be the category of pointed (general) topological spaces. Then one has a functor

$$\mathbf{CW}\colon \mathrm{Top}_* \to \mathrm{Top}_*$$

that associates to each $X \in \mathrm{Top}$ a **CW**-complex in the sense of Whitehead (section B below). This is again an augmented homotopy idempotent functor.

4. Unpointed augmented functors: In fact the **CW**-approximation of a topological space T can be defined in the unpointed category as $|\mathrm{Sing.}X|$, i.e. the realization of the singular complex. One can show that any homotopically idempotent augmented functor $FX \to X$ that preserves weak equivalences, i.e. is equivalent to a simplicial or continuous functor in the unpointed topological category, is either the empty set $\mathbf{F}X = \emptyset$ or weakly equivalent to $|\mathrm{Sing}X|$, i.e. weakly equivalent to the identity. To see why we assume that our functor $\mathbf{F}$ is continuous or simplicial. This is not a strong assumption since we assume that $\mathbf{F}$ is a homotopy functor (1.C.11). It follows directly from idempotency that $\mathbf{F}(*)$ is contractible if it is *not empty.* Now consider the map $X \times * \to X$ as a continuous family of maps of $*$ to X. Since the functor is unpointed we get a family of induced maps $\mathbf{F}(*) \to \mathbf{X}$. This is the adjoint to the simplicial or continuous function on mapping spaces: $\mathrm{map}(*, X) \to \mathrm{map}(\mathbf{F}(*), \mathbf{F}X)$. Now using the coaugmentation one gets that X is a retract of $\mathbf{F}X$. By idempotency one get that X is equivalent to $\mathbf{F}X$ as needed.

5. Universal A-equivalence: Given any space $A \in \mathcal{S}_*$ there is a universal A-equivalence $\mathbf{CW}_A X \to X$ where $\mathbf{CW}_A\colon \mathcal{S}_* \to \mathcal{S}_*$ is an augmented homotopy idempotent functor. It plays a major role in these notes. See (B) below. Examples (1) and (3) are special cases of (5), but example (2) is not.

B. Construction of $\mathbf{CW}_A$, the universal A-equivalence

Given $A \in \mathcal{S}_*$ we construct a continuous (or simplicial, see (1.C.8)), homotopy idempotent augmented functor $\mathbf{CW}_A\colon \mathcal{S}_* \to \mathcal{S}_*$ such that the map $\mathbf{CW}_A X \to X$

is a universal A-equivalence: For any other A-equivalence $T \to X$ there is, up to homotopy, a factorization $\mathbf{CW}_A X \to T \to X$ of the augmentation above which it is, moreover, unique up to homotopy in $\mathcal{S}_*$. It will shortly turn out that this universal A-equivalence is also the (co-)universal map of 'A-cellular spaces' into X, see (D.2.1) below. The construction of $\mathbf{CW}_A X$ is given here as a colimit of a transfinite telescope $\mathbf{CW}_A X = \operatorname{colim}_{\beta<\lambda} \mathbf{C}_A^\beta X = \mathbf{C}_A^\lambda X$ for large enough $\lambda = \lambda(A)$. The construction is very similar to that of $\mathbf{L}^\beta X$ in (1.B.1) above, so we will omit details.

Here is the inductive step: Given a map $l\colon V \to X$ in $\mathcal{S}_*$, we construct a space $\mathbf{C}V$ and derive maps $V \to \mathbf{C}V \xrightarrow{\mathbf{C}(l)} X$. The idea is that $(\mathbf{C}V)^A$ is a 'better approximation' to X^A than V^A.

First define $\overline{\mathbf{C}}V$ as the pointed homotopy pushout in

$$\begin{array}{ccccc} A \rtimes V^A & \xrightarrow{\ell^A} & A \rtimes X^A & & \\ \downarrow & & \downarrow & & \\ V & \longrightarrow & \overline{\mathbf{C}}V & \xrightarrow{\simeq} \mathbf{C}V & \xrightarrow{\mathbf{C}(\ell)} X \end{array}$$

Since the strict pushout of ev and ℓ^A maps into X, so does the homotopy pushout in a natural way. Define $\mathbf{C}V$ as the factorization $\overline{\mathbf{C}}V \stackrel{\simeq}{\rightarrowtail} \mathbf{C}V \to X$ of the given map $\overline{\mathbf{C}}V \to X$ into a trivial cofibration (w.e.) followed by a fibration. Thus $\mathbf{C}V$ is fibrant too. (It seems as if this construction 'feeds' the desired function complex X^A into the inductively existing one, V^A, until eventually the latter becomes equivalent to the former.) We now consider the map $* \to X$ given by $V = *$, the base point, and define $\mathbf{C}^0(*) = *$ and, by induction, $\mathbf{C}^{\beta+1}V = \mathbf{C}(\mathbf{C}^\beta V)$, while for limit ordinal we take the (homotopy) colimit.

B.1 Proposition: *For $\lambda = \lambda(A)$ the map $\mathbf{C}_A^\lambda(*) = \mathbf{CW}_A \to X$ is a universal A-equivalence.*

Proof: First notice that by construction, if a given map $V \to X$ is an A-equivalence, then $V \to \mathbf{C}V$ is a homotopy equivalence. The collection of maps $\mathbf{C}_A^\beta(*) \to X$ gives a map of towers $\{\mathbf{C}_A^\beta(*)\} \to \{X\}$, with a collection of 'homotopy sections' $\{s\}$ in the ladder of mapping spaces:

$$\begin{array}{ccccccccc} (*)^A & \longrightarrow & (\mathbf{C}(*))^A & \longrightarrow & (\mathbf{C}^2(*))^A & \longrightarrow & (\mathbf{C}^\beta(*))^A & \longrightarrow \cdots \longrightarrow X \\ \downarrow & s\nearrow & \downarrow & s\nearrow & \downarrow & s\nearrow & \downarrow & \\ X^A & = & X^A & = & X^A & = & X^A & \end{array}$$

where s is the obvious adjoint to $A \rtimes X^A \to \mathbf{C}(*)$ that comes with the homotopy pushout which defines $\mathbf{C}$· see (D.4) below for a discussion of half-smash product $\rtimes$. By construction the triangles involving s in the ladder commute up to homotopy.

We now need to prove weak equivalence of function complexes. Since we work in $\mathcal{S}_*$, we can consider homotopy groups and we get that the maps induced by $(s_\beta)_{\beta<\lambda}$ and the fibrations l: $C^\beta(*) \to X$ induce in the limit an isomorphism on the homotopy groups of all components $\pi_0 X^A \simeq \pi_0(C_A^\lambda(*))^A$. Hence the result.

B.2 Universality and idempotency of C_A^λ and $\mathbf{CW}_A$. Idempotency is immediate since if for all $\beta \geq 0$ $W \to W'$ is an A-equivalence, then so is $C_A^\beta(W) \to C_A^\beta(W')$.

As for universality it follows by a similar argument to (1.C) above: If $T \to X$ is an A-equivalence, i.e. $T^A \overset{\simeq}{\to} X^A$ is a weak homotopy equivalence, then we have by the above $\mathbf{CW}_A T \overset{\simeq}{\to} \mathbf{CW}_A X$, and thus for any map t: $T \to X$ that is an A-equivalence, the augmentation $\mathbf{CW}_A X \to X$ factors through T via $\mathbf{CW}_A X \to T \to X$. Since the map $\mathbf{CW}_A X \to T$ is equivalent to the map $\mathbf{CW}_A T \to T$, it is determined uniquely (up to homotopy) by $T \to X$.

Thus we have shown:

B.3 Theorem: *The functor $\mathbf{CW}_A$: $\mathcal{S}_* \to \mathcal{S}_*$ is a homotopy idempotent, augmented continuous functor, with $\mathbf{CW}_A X \to X$ universal among all A-equivalences into X.*

C. A common generalization of $\mathbf{L}_f$ and $\mathbf{CW}_A$ and model category structures

The functors $X \to \mathbf{L}_f X$ and $\mathbf{CW}_A X \to X$ are special cases of one relative functor that, for a given cofibration f: $A \hookrightarrow B$, gives a factorization of g: $X \to Y$ into $X \hookrightarrow \mathbf{CL}_f(g) \to Y$, with $\mathbf{CL}_f(X \to *) = \mathbf{L}_f X$ and $\mathbf{CL}_{*\to A}(* \to X) = \mathbf{CW}_A X$. The functor $\mathbf{CL}_f(g)$ can be characterized by the following universal property: Its value on a map $g : X \to Y$ is universal up to homotopy among decomposition, $X \to W \to Y$, such that

$$\mathrm{map}(B, W) \simeq \mathrm{map}(A, W) \times_{\mathrm{map}(A,Y)} \mathrm{map}(B, Y).$$

For any such factorization via a space W with this property one has a map $\mathbf{CL}_f(g) \to W$ that commutes with the obvious maps in sight, and that map is unique up to homotopy.

Similarly we can easily consider the opposite universal property of $\mathbf{CL}_f(g)$; we will not do it.

C.1 CONSTRUCTION OF $\mathbf{CL}_f(g)$. This proceeds as in the cases of $\mathbf{CW}_A X$ and $\mathbf{L}_f X$. Here the inductive step is a bit more involved: Given a map $V \xrightarrow{g} W$ one defines $CL^1(g) \simeq \mathbf{CL}^1_f(g)$ as the homotopy pushout of

$$\begin{array}{ccc}
B \times V^B \coprod_{A \times V^B} A \times \left(V^A \times_{W^A} W^B\right) & \xrightarrow{f^*} & B \times \left(V^A \times_{W^A} W^B\right) \\
\Big\downarrow{\scriptstyle ev} & & \Big\downarrow \\
V & \longrightarrow & \mathbf{CL}^1(g)
\end{array}$$

where ev is obtained by projection and evaluation. This homotopy pushout has a natural map to W. (Again, as in the proof of (B.1) above, we repeatedly add the desired function complex to the inductively given one until after transfinitely many times they agree.)

C.2 LOCALIZING A MODEL CATEGORY STRUCTURE. The extended localization functor $\mathbf{CL}_f$ can be sometimes understood in the context of localizing or colocalizing the usual model category structure on simplicial sets or topological spaces. This is proved in special cases in [N], [Sm-2] and [HH] and it is not really straightforward. Here only a rough outline of what one would like to do, and has been done in special cases, is given:

Given a model category $\mathcal{M}$ and any map $f\colon A \to B$ that is a cofibration in $\mathcal{M}$, we can change the model category in two opposite directions. In one direction we add weak equivalences by enriching the class of trivial cofibration, thus reducing the class of fibration and, in particular, of fibrant objects. In the other direction we use f to enrich the class of trivial fibration, thus reducing the class of cofibration and cofibrant objects.

In more detail one defines $\mathcal{M}_f$, $\mathcal{M}^f$ as being equal to $\mathcal{M}$ as categories, but with a different class of weak equivalences.

$\mathcal{M}_f$ is obtained from $\mathcal{M}$ by adding the cofibration $f\colon A \hookrightarrow B$ to the class of trivial cofibrations and taking the consequences in terms of fibrations. This reduces the class of fibrations to a smaller one that includes only maps $E \to X$ that are fibration in $\mathcal{M}$ and have the homotopy lifting property with respect to *all* cofibration maps in $\mathcal{M}$ that are f-equivalence. Namely fibrations are maps $E \to X$ that satisfy in $\mathcal{M}$ the following equation for any cofibration that is an f-equivalence $V \longrightarrow W$: (This equation is called the Homotopy Right Lifting Property – HRLP.)

(HRLP) $$\mathrm{map}(W, E) \simeq \mathrm{map}(V, E) \times_{\mathrm{map}(V,X)} \mathrm{map}(W, X)$$

In particular, fibrant objects are f-local objects. The classes of trivial fibrations and cofibration do not change. Thus $\mathcal{M}_f$ is obtained from $\mathcal{M}$ by enriching the class of trivial cofibration, cutting down the class of fibration, leaving unchanged the class of cofibrations and trivial fibrations.

NOTICE that in the above one uses a whole *class* of maps to define the new notion of fibration in $\mathcal{M}_f$. It is far from obvious that using this definition one could actually construct a factorization of an arbitrary map as needed in a model category. This is explained in detail in [HH].

Now, the other proposed model category $\mathcal{M}^f$ would, in general, be obtained dually by enriching the class of trivial fibrations to be the class of the fibrations in $\mathcal{M}$ that have the lifting property HRLP as above with respect to the given map f. This includes *all* the fibration in $\mathcal{M}_f$ defined above but also some more fibre maps in $\mathcal{M}$. The class of cofibration is appropriately reduced. The other two classes do not change. Taking the appropriate smaller class of cofibrations here means that only maps that are repeated extensions of the initial object by f are cofibrations: Thus the map with which we started the discussion, $f\colon A \hookrightarrow B$, is the typical cofibration and $\operatorname{cof} f = B \cup CA$ is the typical cofibrant object. Fibrations and trivial cofibrations remain the same as in the initial category $\mathcal{M}$.

NOTICE For a general non-null map f the above proposal for model category on $\mathcal{M}^f$ does not work as it stands [HH].

As far as we now know only $\mathcal{M}_f$ works well for a general f. It is still unknown whether there exists a modified version of the proposed model category $\mathcal{M}^f$ that does satisfy Quillen's or similar axioms in general. Nofech's thesis [No] shows that it works for the map $* \longrightarrow A$, see Section E below.

C.3 FACTORIZATION. Now the point is that, given a map $V \to X$ in $\mathcal{M}$, it can be viewed either as a map in $\mathcal{M}_f$ or as a map in $\mathcal{M}^f$. One would like to understand factorization of a given map in both categories. The relative localization $V \to \mathbf{CL}_f(V, X) \to X$ is a factorization into trivial cofibration followed by a map with RHLP in $\mathcal{M}_f$ –which may or may not be a fibration in $\mathcal{M}_f$. If f is null map then the same factorization is also one into cofibration followed by a trivial fibration. The other factorizations in $\mathcal{M}_f$ and $\mathcal{M}^f$ are the usual ones in $\mathcal{M}$.

The abstract closed model category discussion above can be considered very concretely in Top_* or $\mathcal{S}_*$. In Top_* this leads one to consider non-CW spaces that are still 'cellular' or cofibrant with respect to some 'locally complicated' space. We have not pursued this path. For simplicity we will restrict attention to $\mathcal{S}_*$ and consider only the map $f\colon * \to A$. We would like to develop tools that allow one to prove that a space $X \in \mathcal{S}_*$ is *cofibrant* with respect to $* \to A$ or is A-cellular. This will shed new light on such topics as the Hilton–Milnor decomposition theorem, the James construction and the symmetric products of Thom–Dold. It is also crucial for understanding the deeper properties of $\mathbf{L}_f$ and $\mathbf{P}_A$. In the foregoing section we set some basic notions about those A-cellular spaces. Further discussion is carried on in Chapter 9 below.

D. Closed classes and A-cellular spaces

In this section we discuss certain full sub-categories of $\mathcal{S}_*$ called *closed classes.* The main example of such classes is the class $C^{\cdot}(A)$ of all A-cellular spaces with $\mathbf{CW}_A X \simeq X$ for a given pointed A. Another important example of a closed class is the class of spaces Y that 'map trivially' to all finite-dimensional spaces (i.e. for which the function complex $\mathrm{map}_*(Y, K) \simeq *$ for any finite complex K is trivial) or to any other class of spaces.

D.1 DEFINITION: *A full subcategory of pointed spaces $C^{\cdot} \subset \mathcal{S}_*$ is called 'closed' if it is closed under weak equivalence and arbitrary pointed homotopy colimits: namely for any (pointed) diagram of spaces in $C^{\cdot}$ (i.e. a functor $\underset{\sim}{X}: I \to C^{\cdot}$) the space $\mathrm{hocolim}_* \underset{\sim}{X}$ is also in $C^{\cdot}$.*

We prove several closure theorems for any closed class $C^{\cdot}$. The most important ones are:

1. $C^{\cdot}$ is closed under finite product.
2. If $X \in C^{\cdot}$ and Y is any (unpointed) space, then $X \rtimes Y = (X \times Y)/* \times Y$ is in $C^{\cdot}$.
3. If $F \to E \to B$ is a fibration sequence with B connected and F, E in $C^{\cdot}$, then B is also in $C^{\cdot}$.
4. If $A \to X \xrightarrow{i} X \cup CA$ is any cofibration sequence and A is in $C^{\cdot}$, then so is the homotopy fibre of i.
5. $C^{\cdot}$ is closed under retracts. (A retract can be obtained as an infinite direct limit of the retraction followed by the inclusion.)

D.2 EXAMPLES OF CLOSED CLASSES

D.2.1 THE CLASS $C^{\cdot}(A)$ of A-cellular spaces and the *partial order* $X \ll Y$ on spaces. This is the smallest closed class that contains a given pointed space A. It can be built by a process of transfinite induction by starting with the full subcategory containing the single space A and closing it repeatedly under arbitrary pointed hocolim and weak equivalences. In Section E.5 below we give a 'cellular' description of spaces in up to homotopy equivalence. We refer to members of $C^{\cdot}(A)$ as *A-cellular spaces* or sometimes simply as *A-spaces.* Notice that by (D.8) below the smash product of A with any pointed space is in this class as well as the half-smash with any unpointed space. When Y is X cellular we sometimes denote it by $X \ll Y$. This notation emphasizes that this relation is a (weak) partial order on the class of (pointed) spaces, as can easily be seen, and that the space Y is 'bigger' than X in the sense that it is obtained by assembling several copies of X together along some hocolim scheme. For connected spaces the order does not depend on the choice of base point. Thus $X \ll Y$ iff $\mathcal{C}^{\cdot}(Y) \subseteq \mathcal{C}^{\cdot}(X)$. Put otherwise $X \ll Y$ iff Y can be built by 'assembling together many copies of X' in a pointed fashion.

D.2.2 THE CLASS $\mathcal{E}^{\cdot}(A \xrightarrow{f} B) = \mathcal{E}^{\cdot}(f)$. Here we start with any map (or a class of maps) $f \in \mathcal{S}_*$ of pointed spaces and consider all spaces X such that the induced map on pointed function complexes

$$\mathrm{map}_*(X, A) \to \mathrm{map}_*(X, B)$$

is a (weak) homotopy equivalence of simplicial sets. Since $\mathrm{map}_*(\mathrm{hocolim}_I X_\alpha, A) \simeq \mathrm{holim}_{I^{op}} \mathrm{map}_*(X_\alpha, A)$, it is immediate that $\mathcal{E}^{\cdot}(f)$ is a closed class. If the map f is a null map, we get the class of space with contractible function complex to $A \times B$. Otherwise we don't know many interesting instances of this class which may often be the class of $*$-cellular spaces, i.e. of spaces weakly equivalent to a point. But this class for, say, $B = *$ one-point space is exactly the class of all spaces X with $\mathbf{P}_A X \simeq *$ by elementary fact 1.A.8 (e.9). It turns out that this class is close but not equal to the class of A-cellular spaces, see (3.B) below. It is easy to see that this class contains all the A-cellular spaces using (1.D).

D.2.3 CLASSICAL CW-COMPLEXES. If we consider the above concepts within the category of general topological spaces and take 'weak equivalence' to be a homotopy equivalence, then the smallest closed class that contains the zero-sphere S^0 is the class of all spaces that are homotopy equivalent to CW-complexes.

D.2.4 MILLER SPACES. This is a specially interesting case of (D.2.2) above: the class of spaces X that map trivially to all finite-dimensional spaces K, namely with $\mathrm{map}_*(X, K) \simeq *$. This includes $K(\pi, 1)$ for a finite group π by Miller's theorem. If we replace finite dimensional by p-completion of a finite nilpotent spaces then we get a different closed class that includes for example all connected infinite loop spaces that have torsion fundamental group. [McG, Thm 3]. It is also interesting to replace the finite group π with a compact Lie group G, see [Dw-1, 1.2, 1.3].

D.2.5 E_*-ACYCLIC SPACES. Since pointed homotopy colimits of acyclic spaces with respect to any generalized homology theory are again acyclic, it follows that the class of E_*-acyclic spaces is closed. To check the statement about homotopy colimits of acyclic spaces it is enough e.g. to consider arbitrary wedges and any homotopy coequalizer of two maps $X \rightrightarrows Y$ since by definition any homotopy colimit is a composition of these operations.

D.2.6 n-CONNECTED SPACES. Since any pointed homotopy colimit of n-connected spaces for any fixed integer n is again n-connected, the class of n-connected spaces is closed. In fact this class, when intersected with spaces having the homotopy type of CW-complexes, is simply the class $C^{\cdot}(S^{n+1})$ generated by the $n+1$-sphere. Similarly, the class of spaces with vanishing homology with given coefficient up to a given dimension is also closed.

D.2.7 UNPOINTED HOMOTOPY COLIMITS. We shall see in Chapter 9 that given any closed class of spaces, if we further close it under *unpointed* homotopy colimits indexed by *any contractible* indexing small category, the class will still be closed; in fact it will not change at all.

D.3 POINTED AND UNPOINTED HOMOTOPY COLIMITS. Let A be a pointed space. We have considered $C^{\cdot}(A)$, the smallest class of pointed spaces closed under arbitrary **pointed** hocolim, and homotopy equivalence, which contains the space A. Notice that if we consider classes closed under arbitrary *non-pointed* hocolim, we get only two classes: the empty class and the class of all unpointed spaces. This is true, since a class closed under unpointed hocolim that contains a contractible space, contains all weak homotopy types, since every space is weakly equivalent to the free hocolim of its own simplices (see Appendix HL and 1.F above).

Notice also that if A is not empty, then $C^{\cdot}(A)$ contains the one-point space $* \simeq \text{hocolim}_*(A \to A \to A \to \cdots)$ where all the maps in this infinite telescope are the trivial maps into the base point $* \in X$.

In general, given a pointed I-diagram $\underset{\sim}{X}$ we can consider its homotopy (inverse) limit in either the pointed or unpointed category. By definition, these two homotopy limits have the same (pointed or unpointed) homotopy type. They have in fact the same underlying space. On the other hand, the homotopy colimits of $\underset{\sim}{X}$ will generally have a different homotopy type when taken in the pointed or unpointed category: If $\underset{\sim}{*}$ is the I-diagram of base points in $\underset{\sim}{X}$, then by the very definition of homotopy colimits, pointed and unpointed, we have a *cofibration*, with $NI =$ the classifying space (or the nerve) of the category I:

$$NI \to \text{free-hocolim}\, \underset{\sim}{X} \to \text{pointed-hocolim}\, \underset{\sim}{X}.$$

See [B-K, p. 327 & p. 333].

D.3.1 COROLLARY: *If the classifying space of the indexing category I is contractible, then for any pointed I-diagram $\underset{\sim}{Y}$ we have a homotopy equivalence free-*$\text{hocolim}_I \underset{\sim}{Y} \simeq$ *pointed-*$\text{hocolim}_I \underset{\sim}{Y}$.

Remark: Thus over the usual pushout diagram $\cdot \leftarrow \cdot \rightarrow \cdot$ and over the infinite tower $\cdot \rightarrow \cdot \rightarrow \ldots\ldots$ hocolim takes the same value in the pointed and unpointed categories, but not, e.g., over a discrete group.

Thus for any small category I we have $\mathrm{hocolim}_*\{*\} = \{*\}$, while free-$\underset{I}{\mathrm{hocolim}}\{*\} = BI = NI$ is the nerve (or the classifying space) of I.

D.4 HALF-SMASHES AND PRODUCTS IN CLOSED CLASSES. We now show that a closed class $\mathcal{C}^{\cdot}$ is closed under half-smash with an arbitrary unpointed space (i.e. $\mathcal{C}^{\cdot}$ is an *ideal* in $\mathcal{S}_*$ under the operation $\mathcal{C}^{\cdot} \rightarrow \mathcal{C}^{\cdot} \rtimes Y$), and under internal finite Cartesian products. But first:

D.5 GENERALITIES ABOUT HALF-SMASH. Recall the notation $X \rtimes Y = (X \times Y)/* \times Y$ and $X \ltimes Y = (X \times Y)/X \times *$, where X is pointed and Y is unpointed space. This gives a bifunctor $\mathcal{S}_* \times \mathcal{S} \rightarrow \mathcal{S}_*$. There is another bifunctor $\mathcal{S} \times \mathcal{S}_* \rightarrow \mathcal{S}_*$ given by $\tilde{\mathrm{map}}(Y, X)$ where Y is unpointed and X pointed and where $\tilde{\mathrm{map}}(Y, X)$ is the space of *all* maps equipped with the base point $Y \rightarrow * \rightarrow X$. Thus the underlying space of $\tilde{\mathrm{map}}(Y, X)$ is the same as that of the *free* maps while the underlying space of $X \rtimes Y$ is different in general from that of the base point free product $X \times Y$.

There are obvious adjunction identities:

(i) $\mathrm{map}_*(A \rtimes Y, X) = \mathrm{map}_*(A, \tilde{\mathrm{map}}_*(Y, X))$,

(ii) $\mathrm{map}_*(A \rtimes Y, X) = \mathrm{map}(Y, \mathrm{map}_*(A, X))$.

The first identity (i) says that for each $Y \in \mathcal{S}$ the functor $- \rtimes Y\colon \mathcal{S}_* \rightarrow \mathcal{S}_*$ is left adjoint to $\tilde{\mathrm{map}}(Y, -)$, whereas identity (ii) says that for each $A \in \mathcal{S}_*$ the functor $A \rtimes -\colon \mathcal{S} \rightarrow \mathcal{S}_*$ is left adjoint to $\mathrm{map}_*(A, -)$, where this mapping space is taken as an unpointed space.

In particular, we conclude from the general properties of left adjoints:

D.6 PROPOSITION: *For each $A \in \mathcal{S}_*$ and $Y \in S$, the functors $- \rtimes Y$ and $A \rtimes -$ commute with colimits and hocolimits.*

D.6.1 NOTE: To say that $A \rtimes -\colon S \rightarrow \mathcal{S}_*$ commutes with hocolim involves commuting *pointed* hocolim, i.e. the hocolim in $\mathcal{S}_*$ with *unpointed* hocolim in $\mathcal{S}$.

Explicitly: For any base point free diagram of space $\underset{\sim}{Y}\colon I \rightarrow S$ we have an equivalence:

$$A \rtimes (\text{free-} \underset{I}{\mathrm{hocolim}}\, \underset{\sim}{Y}) \simeq \text{pointed-} \underset{I}{\mathrm{hocolim}_*}(A \rtimes \underset{\sim}{Y}).$$

D.7 LEMMA: *If Y is any unpointed space then for any indexing diagram I the functor $- \rtimes Y\colon \mathcal{S}_* \rightarrow \mathcal{S}_*$ commutes with $\underset{I}{\mathrm{hocolim}}$, and if X is any pointed space the functors $X \rtimes -\colon S \rightarrow \mathcal{S}_*$ and $- \wedge X\colon \mathcal{S}_* \rightarrow \mathcal{S}_*$ commute with $\underset{I}{\mathrm{hocolim}}$.*

Proof: We have just considered $X \rtimes -$. Similarly $- \wedge Y$ is left adjoint to $\mathrm{map}_*(Y, -)$ and again commutes with colim and hocolim.

D.8 THEOREM: *If X is in any closed class $C^{\cdot}$ then:*

1. *For any (unpointed) space Y the half-smash $X \rtimes Y$ is in $C^{\cdot}$.*
2. *For any (pointed) B-cellular space Y and any A-cellular space X the smash $X \wedge Y$ is an $(A \wedge B)$-cellular space.*

D.9 REMARK: Notice that, X being a retract of the half-smash, one can also read (D.8(1)) backward: If $X \rtimes Y$ is in $C^{\cdot}$ then so is X for Y non-empty.

Proof: To prove (1) we start with an example showing that $X \rtimes S^1$ is an X-cellular space. In fact it can be obtained directly as a pointed hocolim of the pushout diagram:

$$\begin{array}{ccc} X \vee X & \xrightarrow{\text{fold}} & X \\ {\scriptstyle\text{fold}}\downarrow & & \downarrow \\ X & \longrightarrow & X \rtimes S^1 \end{array}$$

This diagram is obtained simply by half-smashing X with the diagram that presents S^1 as *free*-hocolim of discrete sets:

$$\begin{array}{ccc} \{0,1\} & \to & \{0\} \\ \downarrow & & \downarrow \\ \{1\} & \to & S^1 \end{array}$$

By induction we present S^{n+1} as a pushout $* \leftarrow S^n \to *$ which gives by induction $X \rtimes S^{n+1}$ as a pushout along $X \leftarrow X \rtimes S^n \to X$, that arises, since (D.6) $(X \rtimes -)$ commutes with free-hocolim on the right (smashed) side. Since the filtration of Y by skeleton $Y_0 \subset Y_1 \subset \cdots$ presents $Y_{n+1} = Y_n \cup (C \coprod S^n)$ we get upon half-smashing with X a presentation of $Y \rtimes X$ as a pointed-hocolim.

D.9.1 REMARK: Here is a 'global' formulation of the above proof using (D.6): Present the space Y as $\text{free-hocolim}_{\Gamma Y}\{*\}$, where ΓY is any small category whose nerve is equivalent to Y, see (1.F) and Appendix HL. Further, by $\underset{\sim}{\{*\}}$ we denote the ΓY-diagram consisting of the one-point space for each object of ΓY. Now by (D.6) above:

$$X \rtimes Y = X \rtimes \text{free-hocolim}_{\Gamma Y}\underset{\sim}{\{*\}} = \text{pointed-hocolim}_{\Gamma Y} X \rtimes \underset{\sim}{\{*\}}.$$

Thus $X \rtimes Y$ is directly presented as a pointed hocolim of a pointed diagram consisting solely of copies of the space X itself.

Now to prove (2) one just notices that $X \wedge Y = (X \rtimes Y)/X \times \{pt\}$, so $X \wedge Y$ is certainly an X-cellular space. Now since any pointed-hocolim commutes with smash-product we work by double induction: First we show by induction on the presentation of X as an A-space that $X \wedge B$ is an $A \wedge B$-space, and then by induction on the presentation of Y as a B-space that $X \wedge Y$ is an $A \wedge B$-space as needed.

D.10 FIBRATIONS AND CLOSED CLASSES. While closed classes are defined using pointed homotopy colimits and, in particular, cofibrations, the more interesting results relate the cellular structures of members in fibration sequences. This is harder to come by since e.g. the homotopy fibre of a map is not easily related by homotopy colimits to the base and total spaces. A deeper look into these matters is taken in Chapter 9. Here we confine ourselves to results needed in the coming developments and a few other examples.

D.11 THEOREM: *Let $F \to E \to B$ be any fibration of pointed spaces with connected B. If F and E are members of some closed class $\mathcal{C}^{\cdot}$ then so is B.*

We shall see later (3.E.1) that this implies:

D.12 COROLLARY: *In a fibration, if the base and total spaces are ΣA-cellular then the fibre is A-cellular.*

The following consequence immediately implies the generalized formulation of a lemma of Zabrodsky and Miller given in [M, 4.6], compare also [B-4, 4.7]:

D.13 COROLLARY: *Let $F \to E \to B$ be any fibration sequence over a connected base B. If both the fibre and the total space have a trivial pointed function complex to a given pointed space Y, then so does the base space B.*

The second corollary follows immediately by observing (D.2.2) that the class of spaces with a trivial function complex to a given space is closed.

Proof: We define a sequence of fibrations $F_i \to E_i \to B$ by $E_0 = E, F_0 = F$, $E_{i+1} = E_i \cup CF_i$ and F_{i+1} is the homotopy fibre of an obvious map $E_{i+1} \to B$. All E_i, F_i are naturally pointed spaces.

$$\begin{array}{ccccc}
F & \longrightarrow & E & \longrightarrow & B \\
 & & \downarrow & & \\
F_1 & \longrightarrow & E \cup CF = E_1 & \longrightarrow & B \\
 & & \downarrow & & \\
F_2 & \longrightarrow & E_1 \cup CF_1 = E_2 & \longrightarrow & B \\
 & & \downarrow & & \\
 & & \cdots & & \\
 & & E_\infty \simeq B & &
\end{array}$$

By Ganea's theorem [G] (but see also Appendix HL at the end) $F_{i+1} \simeq F_i * \Omega B \simeq \Sigma(F_i \wedge \Omega B)$ and therefore connectivity of F_{i+1} is at least i, since F_0 is (-1)-connected. Notice that by the above closure properties of closed classes, since E_0, F_0 are in $C^{\cdot}$ spaces so are E_i, F_i for all i. But since conn $F_i \to \infty$, we deduce that hocolim $E_i = B$. Therefore B is also in $C^{\cdot}$, as needed.

We now turn to the somewhat surprising closure property of closed classes (D.1(4)) that will be treated more fully only in Chapter 9.

D.14 THEOREM: *For any map $A \to X$ of pointed spaces, the homotopy fibre F of $X \to X \cup CA$ satisfies $\mathbf{P}_A F \simeq *$. In particular $\mathbf{P}_A(A) \simeq *$. Moreover, F is A-cellular.*

Outline of Proof: The proof uses the following diagram:

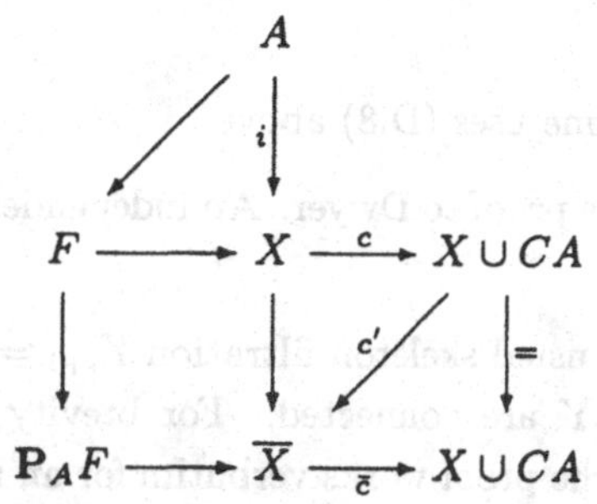

where the vertical arrows are given by the fibrewise localization (1.F.1) of the top row. Thus the fibre map $\bar{c}$ is induced from the composition $X \cup CA \to$Baut $F \to$ Baut $\mathbf{P}_A F$, where by $\mathbf{B}$ we denote here the classifying space functor. Taking F to be the usual path space we have a well defined map $i' : A \to F$ of A to the homotopy fibre. Since map$(A, \mathbf{P}_A F) \simeq *$, by construction of $\mathbf{P}_A F$ the composition $A \to X \to \overline{X}$ factorizing through $\mathbf{P}_A F$ is null-homotopic, where the null homotopy comes from the cone $A \to F \to F \cup CA \to \mathbf{P}_A F$ that defines $\mathbf{P}_A F$. This null homotopy gives a well defined map $c' : X \cup CA \to \overline{X}$ rendering the diagram commutative.

Therefore the fibration $\bar{c}$ is a split fibration having c' as a section. Also, since $F \to \overline{X}$ factors through $X \cup CA$ it is a null homotopic map. But the splitting of $\bar{c}$ implies from the long exact sequence of the fibration that the map $\mathbf{P}_A F \to \overline{X}$ is injective on pointed homotopy class $[W, -]_*$ for any $W \in \mathcal{S}_*$. And since $F \to \mathbf{P}_A F \to \overline{X}$ is null homotopic we conclude that $F \to \mathbf{P}_A F$ is null. Now idempotency of $\mathbf{P}_A$ implies $\mathbf{P}_A F \sim *$ as needed. A complete proof that F is A-cellular is longer and given in Chapter 9 below. See (9.A.10).

D.15 CLOSURE UNDER PRODUCTS. Many of the pleasant properties of $\mathbf{CW}_A$ depend on its commutation with finite products. This commutation rests on the following basic closure property of any closed class.

D.16 **Theorem**: *Any closed class $\mathcal{C}^{\cdot}$ is closed under finite products: If $X, Y \in \mathcal{C}^{\cdot}$ then so is $X \times Y$.*

D.16.1 **Remark**: It is well known that an infinite product of S^1's does not have the homotopy type of a **CW**-complex, i.e. the class of all **CW**-complexes in Top_* in Example D.2.3 above is not closed under arbitrary products. Also notice that an infinite product of $H\mathbb{Z}$-acyclic spaces may not be acyclic, thus the closed class of all $H\mathbb{Z}$-acyclics is not closed under arbitrary products.

D.16.2 **Remark**: If $A = \Sigma A'$ and $B = \Sigma B'$ where $A, B \in \mathcal{C}^{\cdot}$, then $A \times B$ is easily seen to be in $\mathcal{C}^{\cdot}$ via the cofibration

$$A' * B' \to \Sigma A' \vee \Sigma B' \to \Sigma A' \times \Sigma B'.$$

Since $A' * B' \cong \Sigma A' \wedge B'$ one uses (D.8) above.

Proof of D.16: We owe the proof to Dwyer. An independent proof can be extracted from [B-1].

Proof: We filter Y by its usual skeleton filtration $Y_{n+1} = Y_n \cup e^{n+1} \cdots$.

We may assume X, Y are connected. For brevity of notation we add one *pointed* cell at a time but the proof works verbatim for an arbitrary number of cells. Let $P(n)$ be the subspace of $X \times Y$ given by

$$P(n) = \{*\} \times Y \cup X \times Y_n.$$

Clearly the tower $P(n) \hookrightarrow P(n+1)$ is 'cofibrant' and its colimit $X \times Y$ is equivalent to its homotopy colimit. Since $\mathcal{C}^{\cdot}$ is closed under hocolim it is sufficient to show, by induction, that $P(n) \in \mathcal{C}^{\cdot}$ for all $n \geq 0$. For $n = 0$, we have $P(0) = X \vee Y$ clearly in $\mathcal{C}^{\cdot}$. Now $P(n)$ is given as a homotopy pushout diagram:

$$\begin{array}{ccccc}
X \times S^{n-1} \cup * \times D^n & \xrightarrow{\simeq} & X \rtimes S^{n-1} & \longrightarrow & \{*\} \times Y \cup X \times Y_{n-1} \\
\downarrow & & \downarrow & & \downarrow \\
X \times D^n & \xrightarrow{\simeq} & X \rtimes D^n & \longrightarrow & P(n)
\end{array}$$

coming from the presentation of Y_n as a pushout over a *pointed* diagram: $Y_{n-1} \leftarrow S^{n-1} \to D^n$. Since the upper-left corner is equivalent to the half-smash $X \rtimes S^n$, it is in $\mathcal{C}^{\cdot}$ by Lemma (D.8.1) above. Notice that all the maps are pointed. Therefore $P(n)$ is a homotopy pushout of members of $\mathcal{C}^{\cdot}$ as needed.

D.17 Corollary: *For any two A-cellular spaces X, Y their product $X \times Y$ is an A-cellular space.*

Proof: Consider the class $C^{\cdot}(A)$. By the theorem just proved it is closed under finite product, therefore the product of any two A-cellular spaces is A-cellular.

E. A-Homotopy theory and universal properties

In this section we describe some initial elements of A-homotopy theory which is a special case of (C.2) above. This will allow us to better grasp properties of the functor $\mathbf{CW}_A$. In this framework one replaces the usual sphere S^0 in the usual homotopy theory of **CW**-complexes or simplicial sets by an arbitrary space A. It can be considered in the framework of general, compactly generated spaces where A can be chosen to be any such space. We will, however, restrict our discussion to $A \in \mathcal{S}_*$, a pointed space.

It turns out that there is a model category structure on $\mathcal{S}_*$ denoted by $\mathcal{S}_*^A$, where a weak equivalence $f: X \to Y$ is a map that induces a usual weak equivalence

$$\text{map}_*(A, f): \text{map}_*(A, X) \to \text{map}_*(A, Y)$$

of function complexes, and A-fibre maps are defined similarly. Cofibrations are then determined by the lifting property [N]; see (C.2) above.

The cofibrant objects, i.e. the **CW**-complexes, are A-cellular spaces. The natural homotopy groups in this framework are A-homotopy groups

$$\begin{aligned} \pi_i(X; A) &\equiv [\Sigma^i A, X]_* \\ &= \pi_i \ \text{map}_*(A, X, \text{null}) \\ &= [A, \Omega^i X]_*. \end{aligned}$$

The classical Whitehead theorem about **CW**-complexes takes in the present context the form:

E.1 Theorem (A-Whitehead theorem): *A map $f: X \to Y$ between two pointed connected A-cellular spaces has a homotopy inverse (in the usual sense) if and only if it induces a homotopy equivalence on pointed function complexes*

$$(*) \qquad \text{map}_*(A, X) \xrightarrow{\simeq} \text{map}_*(A, Y),$$

or equivalently, iff f induces an isomorphism on the pointed homotopy classes:

$$(**) \qquad [A \rtimes S^n, X]_* \xrightarrow{\cong} [A \rtimes S^n, Y]_*$$

for all $n \geq 0$. If the two pointed function complexes are connected, i.e. $\pi_0(X;A) \simeq \pi_0(Y;A) \simeq *$ or if $A = \Sigma A'$ is a suspension, then a necessary and sufficient condition is that it induces an isomorphism on A-homotopy groups:

$$\pi_*(X;A) \xrightarrow{\cong} \pi_*(Y;A).$$

Proof: It is sufficient to show that under $(*)$, for every $W \in C^{\cdot}(A)$, we have that $\mathrm{map}_*(W,X) \xrightarrow{\simeq} \mathrm{map}(W,Y)$ is a homotopy equivalence. This can be easily shown by a transfinite induction on the presentation of W as a hocolim of spaces in $C^{\cdot}(A)$. Namely, one needs only to show that the class of spaces W for which $\mathrm{map}_*(Y,f)$ is a homotopy equivalence is a closed class. But this is the content of (D.2.2) above. Since by assumption it contains A, it follows that it contains also $C^{\cdot}(A)$ and therefore, by our assumption, it contains both X and Y. Thus we get a homotopy inverse to $X \to Y$ by taking $Y = W$. This shows that in fact $(*)$ implies a homotopy equivalence.

A more careful argument is necessary to show that looking at pointed homotopy classes as given by $(**)$ is a sufficient condition for a weak homotopy equivalence of the function complexes and thus by $(*)$ for f being a weak equivalence. The difficult point is that when one expresses that isomorphism $(*)$ in terms of *pointed* homotopy classes one must keep a fixed map of A to X and Y fixed throughout the homotopy while in $(**)$ we keep only the base point fixed. This was proved in [Ca-R].

E.2 Half-suspensions $\tilde{\Sigma}^n X$. A basic building block for $\mathbf{CW}_A$ is the half-smash $A \rtimes S^n = S^n \times A \cup D^{n+1} \times \{*\}$ with the base point $\{*\} \times \{*\}$. We denote these spaces by $\tilde{\Sigma}^n A$, and call them half n-suspensions of A.

Just as a homotopy class $\alpha \in \pi_n \mathrm{map}_*(A, X; \mathrm{null})$ in the null component is represented by a pointed map $\Sigma^n A \to X$, so does a map $\tilde{\alpha}$: $\tilde{\Sigma}^n A \to X$ represent an element in $\pi_n \mathrm{map}_*(A, X; f)$ of the f-component where f: $A \to X$ is any map. The map f is obtained from $\tilde{\alpha}$ by restricting $\tilde{\alpha}$ to $* \times A \subseteq \tilde{\Sigma}^n A$. Notice that if A itself is a suspension $A = \Sigma B$, then $\tilde{\Sigma}^n A \cong \Sigma^n A \vee A$ [DF-4] *but in general such a decomposition does not hold.* Thus for suspension $A = \Sigma B$, an element $\tilde{\alpha}$ as above is given simply by a pair $(\alpha \vee f)$: $\Sigma^n A \vee A \to X$. In that case, of course, all the components of $\mathrm{map}_*(A, X)$ have the same homotopy type.

E.3 Elementary construction of $\mathbf{CW}_A X$. Let c_0: $C_0 X = V_{\alpha \in I} \tilde{\Sigma}^i A \xrightarrow{\alpha} X$ be the wedge of all the pointed maps $\tilde{\Sigma}^i A \to X$ from all half-suspensions $\tilde{\Sigma}^i A$ to X.

Clearly the map c_0 induces a surjection on the homotopy classes $[\tilde{\Sigma}^i A, -]$ for every $i \geq 0$. We now proceed to add enough 'A-cells' to C_0, so as to get an isomorphism on these classes. We take the first (transfinite) limit ordinal $\lambda = \lambda(A)$ bigger than the cardinality of A itself (= the cardinality of the simplices or cells or points in A).

The ordinal $\lambda = \lambda(A)$ clearly has the *limit property: Given any transfinite tower of spaces of length* λ

$$Y_0 \to Y_i \to \cdots \to Y_n \cdots \to Y_w \to Y_{w+1} \to \cdots \to Y_\alpha \to \cdots \quad (\alpha < \lambda)$$

every map $\tilde{\Sigma}^i A \to \varinjlim_{\alpha<\lambda} Y_\alpha$ *factors through* $\tilde{\Sigma}^i A \to Y_\beta$ *for some ordinal* $\beta < \lambda$.

Proof: This is clear for every individual cell of $\tilde{\Sigma}^i A$, and since the number of these cells is strictly smaller than the cofinality of λ, it is true for $\tilde{\Sigma}^i A$. We proceed to construct a λ-tower of correction $C_0 = C_0 X \to C_1 X \to C_2 X \to \cdots \to C_\beta X \to \cdots$ to our original map $C_0 \to X$:

(E.4)

$$\begin{array}{ccccc}
D_0 = \bigvee_{K_0} \tilde{\Sigma}^i A & & D_1 = \bigvee_{K_1} \tilde{\Sigma}^i A & & D_\beta = \bigvee_{K_\beta} \tilde{\Sigma}^i A \cdots \\
\downarrow{\scriptstyle k_0} & & \downarrow & & \downarrow{\scriptstyle k_\beta} \\
C_0 = \bigvee_{I_0} \tilde{\Sigma}^i A & \longrightarrow & C_1 X \cdots & \longrightarrow & C_\beta X \longrightarrow \cdots (\beta \leq \lambda) \\
\downarrow{\scriptstyle c_0} & & \downarrow & & \downarrow{\scriptstyle c_\beta} \\
X = & & X = & & X = \cdots
\end{array}$$

Since $C_0 \to X$ is surjective on the A-homotopy of all components of $\mathrm{map}_*(A, X)$, we proceed to kill the kernel in a functorial fashion. In order to preserve functoriality we kill it over and over again: First notice that any element $\tilde{\alpha}$: $\tilde{\Sigma}^n A \to X$ representing an A-homotopy class in the component $\tilde{\alpha}|\{*\} \times A = f$: $A \to X$ is null homotopic *in that component* iff $\tilde{\alpha}$ can be extended along the map

(E.4.1) $$\tilde{\Sigma}^n A = S^n \times A \cup D^{n+1} \times \{*\} \hookrightarrow D^{n+1} \times A.$$

Now let k_0: $D_0 \to C_0$ be the wedge of all maps g: $\tilde{\Sigma}^i A \to C_0$ *with a given* extension as (E.4.1) of $c_0 \circ g$ (the space D_0 being a point if there are no such extensions). Thus $D_0 \to C_0$ captures every null homotopic map $\tilde{\Sigma}^i A \to C_0 \to X$ many times. The map $D_0 \to C_0$ is given by g. We define C_1X as the pushout along the extension to $D^{n+1} \times A$:

$$\begin{array}{ccc} \bigvee_{K_0} \tilde{\Sigma}^i A & \to & \bigvee_{K_0} D^{n+1} \times A \\ \downarrow & & \downarrow \\ C_0 & \longrightarrow & C_1 = C_1X \end{array}$$

In this fashion we proceed by induction. The map $C_1X \to X$ is given by the null homotopies in the indexing set of $D_0 = \bigvee_{k_0} \tilde{\Sigma}^i A$. Taking limits at limits ordinal we define a functorial tower $C_\beta X$ for $\beta \leq \lambda$. We now define $\mathbf{CW}_A X = C_\lambda X$. This is the classical small object argument [Q-1].

Since c_0 induces a surjection on A-homotopy sets $[\tilde{\Sigma}^i A, X]$ for $i \geq 0$ on all components we get immediately that so does c_β for all $\beta \leq \lambda$. The limit property of $\lambda = \lambda(A)$ now easily implies that $C_\lambda X \to X$ is injective in $\pi_i(\ , A; f)$ for any $f\colon A \to X$. Since every null homotopic composition $\tilde{\Sigma}^i A \to C_\lambda X \to X$ factors through $\tilde{\Sigma}^i A \to C_\beta X \to X$ for some β, a composition that is also null homotopic by commutativity, therefore this map is null homotopic in $C_{\beta+1}X$ and thus in $C_\lambda X$, as needed.

E.5 A SMALLER NON-FUNCTORIAL A-CELLULAR APPROXIMATION can be built by choosing representatives in the associated homotopy classes. But it is clear that, in general, even if A, X are of finite complexes $\mathbf{CW}_A X$ may not be of finite type, since $\mathbf{CW}_{S^2}(S^1 \bigvee S^n) \approx \bigvee_\infty S^n$ and this construction is just the universal cover of $S^1 \bigvee S^n$.

E.6 COROLLARY: *Let A be a finite complex. Then for any countable space X we have the following form:*

$$\mathbf{CW}_A X = (\bigvee \tilde{\Sigma}^i A) \cup_{\varphi_1} C\tilde{\Sigma}^{i_1} A \cup_{\varphi_2} C\tilde{\Sigma}^{i_2} A$$
$$\cdots \cup_{\varphi_\ell} C\tilde{\Sigma}^{i_\ell} A \cup \cdots$$

where the 'characteristic maps' φ_ℓ are defined over $\tilde{\Sigma}^{i_\ell} A$ for $0 \leq \ell < \infty$, and therefore $\mathbf{CW}_A X$ is also a countable cell complex.

E.7 REMARK If A is a *finite* suspension space $A = \Sigma B$ of *pointed* B, we have $\tilde{\Sigma}^i A \simeq \Sigma^i A \vee A$ and therefore, in order to kill the kernels of $C_\beta \to X$, it is sufficient to attach cones over the usual suspension of A: $\Sigma^i A \to C_\beta$. Thus in this case the A-cellular approximation to X has the usual form

$$\mathbf{CW}_A X = \left(\bigvee \Sigma^{i_1} A\right) \cup_{\varphi_1} C\Sigma^{i_2} A \cup_{\varphi_2} C \cdots \Sigma^{i_2} A \cdots$$

which is just the usual **CW**-complex for $A = S^1 = \Sigma S^0$, and where X is any connected **CW** complex.

As is usual in homotopy theory, any map $X \to Y$ can be turned into a cofibration $X \hookrightarrow X' \to Y$ where $X \hookrightarrow X'$ is an A-cofibration, i.e. X' is obtained from X by adding 'A-cells' and $X' \to Y$ is a trivial fibration, i.e. in particular it induces an isomorphism on A-homotopy groups (compare (C.2) above). Thus if $Y \simeq *$ we get $X' \simeq \mathbf{P}_A X$, since $\mathrm{map}_*(A, \mathbf{P}_A X) \simeq \mathrm{map}_*(A, *)$ and $X \longrightarrow \mathbf{P}_A X$ is an A-cofibration.

If, on the other hand, we take $X \simeq *$, the factorization becomes $* \to \mathbf{CW}_A Y \to Y$ where $\mathbf{CW}_A Y$ now appears as the A-cellular approximation to X with the same A-homotopy in all dimensions.

E.8 UNIVERSAL PROPERTIES.

We now show that $r\colon \mathbf{CW}_A X \to X$ has two universal properties:

(U1) [B-2, 7.5] The map r is initial among all maps $f\colon Y \to X$ with $\mathrm{map}_*(A, f)$ a homotopy equivalence. Namely for any such map there is a factorization $\tilde{f}$:

$$\begin{array}{ccc} \mathbf{CW}_A X & \xrightarrow{r} & X \\ {\scriptstyle \tilde{f}}\big\downarrow & \nearrow & \\ Y & & \end{array}$$

and such $\tilde{f}$ with $f \circ \tilde{f} \sim r$ is unique up to homotopy.

(U2) The map r is terminal among all map $\omega\colon W \to X$ of spaces $W \in \mathcal{C}^{\cdot}(A)$ into X. Namely for every ω there is a $\tilde{\omega}\colon W \to \mathbf{CW}_A X$ with $r \circ \tilde{\omega} \sim \omega$ unique up to homotopy.

Proof: Both (U1) and (U2) are easy consequences of the functoriality of $\mathbf{CW}_A$ when coupled with the A-Whitehead theorem. Thus to prove (U1) consider $\mathbf{CW}_A(f)\colon \mathbf{CW}_A Y \to \mathbf{CW}_A X$. This map is an A-equivalence between two A-cellular spaces, therefore it is a homotopy equivalence. Uniqueness follows by a simple diagram chase using naturality and idempotency of $\mathbf{CW}_A$. To prove (U2): One gets a map $A \to \mathbf{CW}_A X$ by noticing that $\mathbf{CW}_A W \simeq W$, so $\mathbf{CW}(\omega)$ gives the unique factorization. Furthermore, uniqueness of factorization implies that each one of these universal properties determines $\mathbf{CW}_A X$ up to an equivalence which itself is unique up to homotopy. This proves (U1) and (U2).

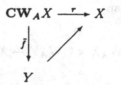

E.9 PROPOSITION: *The following conditions on pointed spaces A and B are equivalent:*

(1) *For any space X there is an equivalence $\mathbf{CW}_A X \simeq \mathbf{CW}_B X$.*
(2) $C^{\cdot}(A) = C^{\cdot}(B)$.
(3) *A map $f\colon X \to Y$ is an A-equivalence if and only if it is a B-equivalence.*
(4) $A \simeq \mathbf{CW}_B A$ *and* $B \simeq \mathbf{CW}_A B$.

Proof: These equivalences follow easily from the universal properties of $\mathbf{CW}_A X \to X$.

(1)⇔(2) Since the members of $C^{\cdot}(A)$ are precisely the space X for which $\mathbf{CW}_A X \simeq X$, this is clear from universality.

(1)⇔(3) Clearly $\operatorname{map}(B, f)$ is an equivalence and $\mathbf{CW}_B f$ is a homotopy equivalence. But since by (1)⇔(2), $\mathbf{CW}_B f \simeq \mathbf{CW}_A f$ and we get (3).

(2)⇔(4) This is immediate from the definitions.

E.10 THEOREM: *For any $A, X, Y \in \mathcal{S}$ there is a homotopy equivalence*

$$\Psi\colon \mathbf{CW}_A(X \times Y) \to \mathbf{CW}_A X \times \mathbf{CW}_A Y.$$

Proof: There is an obvious map

$$g\colon \mathbf{CW}_A X \times \mathbf{CW}_A Y \to X \times Y.$$

It is clear that g induces a homotopy equivalence $\operatorname{map}(A, g)$ and therefore the map Ψ in the theorem induces the same equivalence $\operatorname{map}(A, \Psi)$. But by Corollary D.17 the range of Ψ is an A-cellular space. Thus by the A-Whitehead theorem Ψ is a homotopy equivalence.

E.11 LEMMA: *If $X \simeq \mathbf{CW}_A X$ and Y is a retract of X, then $Y \simeq \mathbf{CW}_A Y$.*

Proof: The retraction $r : X \leftrightarrows Y$ implies that the map $\mathbf{CW}_A Y \to Y$ is a retract of the homotopy equivalence $\mathbf{CW}_A X \to X$. But a retract of an equivalence is an equivalence.

3. COMMUTATION RULES FOR Ω, $\mathbf{L}_f$ AND $\mathbf{CW}_A$, PRESERVATION OF FIBRATIONS AND COFIBRATIONS

Introduction

In this chapter we begin to consider the behaviour of fibration and cofibration under $\mathbf{L}_f$ and $\mathbf{CW}_A$. As in the case of localization at a subring of the rationals or p-completion, the main technical property of both $\mathbf{L}_f$ and $\mathbf{CW}_A$ is that, under some relatively mild restrictions, they 'nearly' preserve fibration sequences. In particular, they nearly commute with the loop functor: We first prove the equivalences $\mathbf{L}_f\Omega X \simeq \Omega\mathbf{L}_{\Sigma f}X$ and $\mathbf{CW}_A\Omega X \simeq \Omega\mathbf{CW}_{\Sigma A}X$ (notice Σf) and draw a few quick consequences. We then consider several cases where $\mathbf{L}_f$ and $\mathbf{CW}_A$ preserve fibration sequence. The most general results are obtained in Chapter 5 below. Here we continue by relating $\mathbf{CW}_A$ and $\mathbf{P}_A$ via a fibration sequence $\mathbf{CW}_A X \longrightarrow X \longrightarrow \mathbf{P}_{\Sigma A}$. We show that this is in fact a fibration sequence whenever the composite is null homotopic: This in particular implies that if $\mathbf{P}_{\Sigma A}X \simeq *$, then X is A-cellular for any $A, X \in S_*$.

In section 7.C below we give several applications showing that certain naturally arising spaces have A-cellular structure, e.g. certain E_*-acyclic spaces can be built from finite ones.

A. Commutation with the loop functor

In the first section we prove that both $\mathbf{L}_f$ and $\mathbf{CW}_A$ for arbitrary $f, A \in S_*$ 'commute' with the loop functor in the following sense (compare [B-4]):

A.1 THEOREM: *Let $f\colon A \longrightarrow B$ be any map in S_* and $X \in S_*$ a connected space. There is a natural homotopy equivalence*

$$\mathbf{L}_f\Omega X \simeq \Omega\mathbf{L}_{\Sigma f}X.$$

A.2 THEOREM: *Let $A, X \in S_*$ be pointed and connected spaces. There is a natural homotopy equivalence*

$$\mathbf{CW}_A\Omega X \simeq \Omega\mathbf{CW}_{\Sigma A}X.$$

Remark: Thus both $\mathbf{L}_f\Omega X$ and $\mathbf{CW}_A\Omega X$ have a natural loop space structure where the corresponding classifying spaces are $\mathbf{L}_{\Sigma f}X$ and $\mathbf{CW}_{\Sigma A}X$. By induction we get a similar result for $\mathbf{L}_f\Omega^n X$ and $\mathbf{CW}_A\Omega^n X$. Still, it is not known whether for any other $W \in S_*$ the space $\mathbf{L}_f\,\mathrm{map}_*(W, X)$ is a W-function space, i.e. has the form $\mathrm{map}_*(W, Y)$ for some Y. Certainly one cannot simply take $Y = \mathbf{L}_{W\wedge f}X$.

A.3 LEMMA: *If X is a connected pointed space, then there is a natural loop space structure on $\mathbf{L}_f\Omega X$ with respect to which the coaugmentation map $\Omega X \longrightarrow \mathbf{L}_f\Omega X$ is homotopic to a loop map.*

Remark: The lemma says in other words that there is a space Y with $\Omega Y \simeq \mathbf{L}_f\Omega X$, and this space Y is natural in X and comes with a loop map $\Omega X \longrightarrow \Omega Y$. Notice that the proof only uses two properties of $\mathbf{L}_f$, namely $\mathbf{L}_f$ preserves products up to homotopy equivalence and is coaugmented.

Proof: To equip $\mathbf{L}_f\Omega X$ with a loop structure one uses Segal's 'machine' to identify loop spaces. According to this machine [S], [A], [Pu] a pointed space V is a loop space if it can be embedded in a 'special' simplicial space $W_* = \{W_i\}_{i\geq 0}$ with $W_0 \simeq *$ and $W_1 \simeq V$ with the crucial condition being: for any n the space W_n is homotopy equivalent to the n-th power $W_1^n = W_1 \times W_1 \times \cdots \times W_1$ of the space in dimension one W_1 where the equivalence is given via the structure maps. Namely, $W_n \simeq W_1^n \simeq V^n$ for $n \geq 1$, and where these homotopy equivalences $W_n \longrightarrow W_1^n$ are given as a product of maps $\prod_{i=1}^n \lambda_i$, where $\lambda_i\colon (0,1) \longrightarrow (0,\ldots,n)$ is given by $\lambda_i(0) = i-1$ and $\lambda_i(1) = i$; the usual simplicial operators map Δ_n^{op} to Δ_1^{op}. Together, they give maps $W_n \longrightarrow W_1$ that combine to form an equivalence

$$\prod W(\lambda_i)\colon W_n \longrightarrow W_1 \times \cdots \times W_1.$$

If such a simplicial space (i.e. a functor $\Delta^{\mathrm{op}} \dashrightarrow S_*$) induces a group structure on $\pi_0 W_1$, then Segal's theorem says that there is a natural homotopy equivalence $\Omega|W_*| \simeq W_1$ where $|W_*|$ is the usual realization of W_* (i.e. its homotopy colimit as a diagram of spaces).

Now, given X we present ΩX as a 'Segal loop space' by taking a monoidic version of ΩX, say $\bar{\Omega} X$, the Moore loop space of maps $[0,a] \to X$, so that we get a precise simplicial space made out of the monoid operations and projections into factors. Let the simplicial space $G_*\colon \Delta^{\mathrm{op}} \longrightarrow \mathrm{Top}$ be given by $G_n = (\bar{\Omega} X)^n$; this has the right properties so that $\Omega|G_*| \simeq G_1 = \bar{\Omega} X$. Now the idea is to apply the functor $\mathbf{L}_f$ to the diagram of spaces given by G_* (=a simplicial space). Since $\mathbf{L}_f$ is a functor we still get a simplicial space and since $\mathbf{L}_f$ commutes with finite products up to homotopy it is easy to see that one still gets a 'special' simplicial space, that gives the desired loop space structure on the space $\mathbf{L}_f\Omega X$. In more details, take the simplicial space $\mathbf{L}_f G_*$ with $(\mathbf{L}_f G_*)_n = \mathbf{L}_f(G_n)$, the localization of the product $(\bar{\Omega} X)^n$. Since the homotopy equivalence

$$G_n = (\bar{\Omega} X)^n \longrightarrow (G_1)^n = (\bar{\Omega} X)^n$$

is in this case the identity map given by a product of projection, this same projection λ_i gives on $\mathbf{L}_f G_*$ a product map $\mathbf{L}_f((\bar{\Omega}X)^n) \longrightarrow (\mathbf{L}_f\bar{\Omega}X)^n$, which we know by elementary fact (1.A.8) (e.4) above to be a homotopy equivalence. Therefore $\mathbf{L}_f G_*$ is a special simplicial space, satisfying the said conditions.

Notice that in dimension one we have $\pi_0\mathbf{L}_f G_* = \pi_0\mathbf{L}_f\bar{\Omega}X$, which is clearly a group, since the equivalence $\mathbf{L}_f(X\times X)\simeq \mathbf{L}_f X\times\mathbf{L}_f X$ is natural and since $\mathbf{L}_f$ does not change the set of components when applied to a non-connected space because f is a map of connected spaces. Therefore $\Omega|\mathbf{L}_f G_*|\simeq \mathbf{L}_f G_1\simeq \mathbf{L}_f\bar{\Omega}X$, presenting $\mathbf{L}_f\bar{\Omega}X$ as a loop space. In fact, since $G_*\longrightarrow \mathbf{L}_f G_*$ is a simplicial map between simplicial spaces, it induces $|G_*|\longrightarrow |\mathbf{L}_f G_*|$, a map which gives a loop map

$$\bar{\Omega}X \xleftarrow{\simeq} \Omega|G_*| \longrightarrow \Omega|\mathbf{L}_f G_*| \xrightarrow{\simeq} \mathbf{L}_f\bar{\Omega}X.$$

This equivalence is of loop spaces, so they combine to show that $\bar{\Omega}X\longrightarrow \mathbf{L}_f\bar{\Omega}X$ is a map of loop spaces.

A.4 PROOF OF A.1: We will construct two maps $\ell: \mathbf{L}_f\Omega X \overset{\longleftarrow}{\longrightarrow} \Omega\mathbf{L}_{\Sigma f}X$; each will be given by the appropriate universality conditions on localizations. The uniqueness part of universality will then imply that they are homotopy inverses.

First notice that by elementary fact (e.7) in (1.A.8) above, the map $\Omega X\longrightarrow \Omega\mathbf{L}_{\Sigma f}X$, which is the loop of augmentation $X\longrightarrow \mathbf{L}_{\Sigma f}X$, is a map of ΩX to an f-local space. Therefore, by universality, it factors uniquely up to homotopy through a map l, $\Omega X\xrightarrow{j}\mathbf{L}_f\Omega X\xrightarrow{l}\Omega\mathbf{L}_{\Sigma f}X$.

To construct the map the other way we may proceed by using the result above that $\Omega X\longrightarrow \mathbf{L}_f X$ is a loop map, and so we can classify it to get $X\longrightarrow \overline{W}\Omega X\longrightarrow \overline{W}\mathbf{L}_f\Omega X$. In fact we may use $|\mathbf{L}_f G_*|$ as a model for $\overline{W}\mathbf{L}_f\Omega X$. Then the map in the other direction will be given as a loop map if we construct a map of spaces $\mathbf{L}_{\Sigma f}X\longrightarrow \overline{W}\mathbf{L}_f\Omega X$, by looping down and composing with the obvious maps.

Again we use elementary fact (1.A.8) (e.7) to notice that $\overline{W}\mathbf{L}_f\Omega X$ is Σf-local, and therefore by universality of $\mathbf{L}_{\Sigma f}$ it is sufficient to find a map $X\longrightarrow \overline{W}\mathbf{L}_f\Omega X$. This can be taken as the composite that was constructed above:

$$X\longrightarrow \overline{W}\Omega X\xrightarrow{\overline{W}\ell}\overline{W}\mathbf{L}_f\Omega X.$$

Now we have a diagram of maps

$$\begin{array}{ccc} \Omega X & \xrightarrow{=} & \Omega X \\ \Big\downarrow{\scriptstyle j_1} & & \Big\downarrow{\scriptstyle \Omega j_2} \\ l\colon \mathbf{L}_f \Omega X & \rightleftarrows & \Omega \mathbf{L}_{\Sigma f} X\colon r \end{array}$$

which commutes up to homotopy on both sides, since the bottom arrows were found by universality. This means $r \circ \Omega j_2 \sim j_1$ or $r \circ \ell \circ j_1 \sim j_1$, but by uniqueness of factorization through the universal $\mathbf{L}_f \Omega X$ we get that $r \circ \ell$ is homotopic to the identity. Similarly one gets $\ell \circ r \sim \mathrm{Id}$. This completes the proof. ∎

A.5 PROOF OF (A.1): The proof proceeds in line with(A.3), (A.4), since $\mathbf{CW}_A$ preserves products up to homotopy equivalence, except that here we have augmented rather than coaugmented functors.

So since $\mathbf{CW}_A$ preserves products we can use the same argument as in (A.3) above to conclude that $\mathbf{CW}_A \Omega X$ is still a loop space and the natural map $\mathbf{CW}_A \Omega X \longrightarrow \Omega X$ can be taken to be a loop map.

Now that we have shown that $\mathbf{CW}_A \Omega X$ has loop space structure compatible with that on ΩX we can use universality properties of $\mathbf{CW}_A$ to get the desired equivalence. First take

$$\Omega j_\Sigma\colon \Omega \mathbf{CW}_{\Sigma A} X \longrightarrow \Omega X$$

to be the loop of the structure map for $\mathbf{CW}_{\Sigma A}$.

We consider the factorization of the last map that gives us r:

(A.5.1)
$$\begin{array}{ccc} CW_A \Omega X & \xrightarrow{=} & CW_A \Omega X \\ \Big\downarrow{\scriptstyle r} & & \Big\downarrow{\scriptstyle aug} \\ \Omega CW_{\Sigma A} X & \xrightarrow{\Omega j_\Sigma} & \Omega X \end{array}$$

To get the factorization first note that (Ωj_Σ) induces a homotopy equivalence on $\mathrm{map}_*(A, -)$. Therefore, by universality of the map $(j\Omega)$ we get the map r in (A.5.1) which is unique up to homotopy. To get the map ℓ in the opposite direction we first construct a map $\ell\prime = \text{‘}\overline{W}\ell\text{’}$

$$\ell'\colon \mathbf{CW}_{\Sigma A} X \longrightarrow \overline{W} \mathbf{CW}_A \Omega X$$

where $\overline{W}$ is the classifying functor otherwise denoted by $B-$. Here we use crucially the fact proven above that $\mathbf{CW}_A\Omega X \longrightarrow \Omega X$ is a loop map. One deloops this map to get a map $\overline{W}j\Omega\colon \overline{W}\mathbf{CW}_A\Omega X \longrightarrow X$. We lift the structure map $\mathbf{CW}_{\Sigma A}X \longrightarrow X$ across $\overline{W}j\Omega$ to $\overline{W}\mathbf{CW}_A\Omega X$ to get the desired map ℓ. Again this lift exists by universality of $\mathbf{CW}_{\Sigma A}X$ (E.8, U.1) since $\overline{W}(j\Omega)$ is easily seen by adjunction to induce homotopy equivalence on the pointed mapping space from A: i.e. $\mathrm{map}_*(A, \overline{W}j\Omega)$ is a homotopy equivalence. Since these two maps were defined by universality, it is easily checked as in (A.4) above that these are mutual inverses up to homotopy. This completes the proof of (A.2).

B. Relations between $\mathbf{CW}_A$ and $\mathbf{P}_A$

In this section we prove several results that put $\mathbf{CW}_AX$ and $\mathbf{P}_AX$ in an 'ideal' relation to each other. Intuitively $\mathbf{CW}_AX$ contains all the 'A-information' on X available via the function complex $\mathrm{map}_*(A,X)$ while $\mathbf{P}_AX$ contains what remains of X after all that A-information was destroyed by the nullification functor. Thus $\mathbf{CW}_AX \longrightarrow X$ should morally be the homotopy fibre of $X \longrightarrow \mathbf{P}_AX$. This is 'almost' the case but, as we shall see, not precisely.

B.1 PROPOSITION: *For all $A, X \in \mathcal{S}_*$ one has $\mathbf{P}_A\mathbf{CW}_AX \simeq *$ and $\mathbf{CW}_A\mathbf{P}_AX \simeq *$.*

Proof: The second equivalence is clear since $\mathrm{map}_*(A, \mathbf{P}_AX) \simeq *$ and so $* \longrightarrow \mathbf{P}_AX$ is an A-equivalence, inducing a homotopy equivalence on $\mathrm{map}(A,-)$. Thus $\mathbf{CW}_A$ turns that map into a homotopy equivalence. To see the first equivalence notice that $\mathbf{CW}_AX$ is built out of the space $A \in \mathcal{S}_*$ via a sequence of pointed homotopy colimits. Now in view of the commutativity relations (1.D) with homotopy colimits, if $\mathbf{P}_AX(\alpha) \simeq *$ for each $\alpha \in I$ then $\mathbf{P}_A \,\mathrm{hocolim}_*\, X(\alpha) \simeq \mathbf{P}_A \,\mathrm{hocolim}_*\, \mathbf{P}_AX(\alpha) \simeq \mathbf{P}_A \,\mathrm{hocolim}_*(pt) \simeq \mathbf{P}_A(pt) \simeq (pt)$ where (pt) is the one point space. Thus for any closed class $\mathcal{C}^{\cdot}(A)$ for $A \in \mathcal{S}_*$ one has $\mathbf{P}_AW \simeq *$ for all $W \in \mathcal{C}^{\cdot}(A)$. But $\mathbf{CW}_AX \in \mathcal{C}^{\cdot}(A)$ hence the conclusion $\mathbf{P}_A\mathbf{CW}_A \simeq *$. ∎

We now formulate the main relation between the nullification and cellularization with respect to A.

B.2 THEOREM: *Consider the sequence*

$$\mathbf{CW}_AX \xrightarrow{\ell} X \xrightarrow{r} \mathbf{P}_{\Sigma A}X$$

for arbitrary pointed connected spaces A, X. This sequence is a fibration sequence if (and only if) the composition $r \circ \ell$ is null homotopic. Moreover, the same conclusion holds if $[A,X] \simeq *$.

Remark: Notice that if $\mathbf{CW}_AX \simeq *$, namely $\mathrm{map}_*(A,X) \simeq *$, the conclusion of the theorem is obvious since in that case X is also ΣA-null, so the map r is an equivalence. In examples (C.9) and (C.11) below we will see that the composition is not always null nor is the sequence always a fibration sequence.

Proof: We first prove the theorem under the special assumption: $\mathbf{P}_{\Sigma A}\mathbf{X} \simeq *$, which of course is a special case of the theorem.

B.3 PROPOSITION: *For any two connected* **CW**-*complexes* A, X *in* $\mathcal{S}_*$, *if* $\mathbf{P}_{\Sigma A}\mathbf{X} \simeq *$ *then* $\mathbf{CW}_A X \xrightarrow{\simeq} X$ *is a homotopy equivalence.*

Proof: In the following fibre sequence one shows that the homotopy fibre F must be contractible. Since we assume that X is connected, this implies that the augmentation is an equivalence.

$$\Omega X \longrightarrow F \dashrightarrow \mathbf{CW}_A X \longrightarrow X. \tag{B.3.1}$$

In order to show that $F \simeq *$ one proves:
(1) $\mathrm{map}_*(A, F) \simeq *$,
(2) $\mathbf{P}_A F \simeq *$.

Clearly any space Y that satisfies (1), i.e. is A-null, does not change under $\mathbf{P}_A$, thus (1) and (2) imply $F \simeq *$.

The fibration (B.3.1) implies that $\mathrm{map}_*(A, F)$ is the homotopy fibre of $\mathrm{map}_*(A, \mathbf{CW}_A X) \longrightarrow \mathrm{map}_*(A, X)$ over the trivial component. But by the definition of $\mathbf{CW}_A$ the latter map is a homotopy equivalence, thus its fibre is contractible and (1) holds. To prove (2) we use Theorem 1.H.1 above, with respect to the fibration sequence $\Omega X \longrightarrow F \longrightarrow \mathbf{CW}_A X$. Since both X and A are connected so is $\mathbf{CW}_A X$. First notice that by (3.A.1) $\mathbf{P}_A \Omega \mathbf{X} \simeq \Omega \mathbf{P}_{\Sigma A} X$ which is, by our assumption, contractible. But now Theorem 1.H.1 means that $\mathbf{P}_A F \xrightarrow{\cong} \mathbf{P}_A \mathbf{CW}_A \mathbf{X}$ is a homotopy equivalence. Theorem B.1 above now implies $\mathbf{P}_A F \simeq *$, as claimed in (2). This completes the proof of the proposition. ∎

We now proceed with the proof of Theorem B.2. Let Y be the fibre of $X \longrightarrow \mathbf{P}_{\Sigma A} X$. By Theorem 1.H.2 we deduce that $\mathbf{P}_{\Sigma A}\mathbf{Y} \simeq *$ and therefore, by the proposition just proved, we deduce $\mathbf{CW}_A Y \xrightarrow{\simeq} Y$ is a homotopy equivalence. The following claim now completes the proof:

CLAIM: $\mathbf{CW}_A Y \simeq \mathbf{CW}_A X$.

Proof: The map $Y \longrightarrow X$ gives us a map $Y \simeq \mathbf{CW}_A Y \longrightarrow \mathbf{CW}_A X$. Since both spaces are A-cellular, it suffices by the A-Whitehead theorem (2.E.1) to prove that we have a homotopy equivalence:

$$\mathrm{map}_*(A, \mathbf{CW}_A Y) \simeq \mathrm{map}_*(A, \mathbf{CW}_A X).$$

Since for any space W the map $\mathbf{CW}_A W \longrightarrow W$ is a natural A-equivalence, by the universal property of $\mathbf{CW}_A$ (E.8.U.1), it suffices to show that $\operatorname{map}_*(A, Y) \longrightarrow \operatorname{map}_*(A, X)$ is a homotopy equivalence of function complexes. Consider first the set of components: By definition of Y as a fibre we have an exact sequence of pointed sets:

$$\text{(B.3.2)} \qquad [\Sigma A, \mathbf{P}_{\Sigma A}\mathbf{X}] \longrightarrow [A, \mathbf{Y}] \longrightarrow [A, \mathbf{X}] \longrightarrow [A, \mathbf{P}_{\Sigma A}\mathbf{X}].$$

Now we claim that it follows from the assumption $r \circ \ell \simeq *$ in our theorem, that the right-most arrow is null. This is because, by the universal property of $\mathbf{CW}_A$ (Theorem E.8 (U.2)), every map $A \longrightarrow X$ factors (uniquely up to homotopy) through $\mathbf{CW}_A X \longrightarrow X$, therefore by the assumption of the theorem ($r \circ \ell \simeq *$) its composition with r: $A \longrightarrow X \xrightarrow{r} \mathbf{P}_{\Sigma A}\mathbf{X}$ must be null homotopic.

Now consider the pull-back sequence:

$$\begin{array}{ccc} \operatorname{map}_*(A, Y) & \longrightarrow & \operatorname{map}_*(A, X) \\ \downarrow & & \downarrow \bar{r} \\ * & \longrightarrow & \operatorname{map}_*(A, \mathbf{P}_{\Sigma A}\mathbf{X}; \text{null}) \simeq * \end{array}$$

We just saw that $\bar{r} = \operatorname{map}_*(A, r)$ carries the whole function complex to the null component of $\operatorname{map}_*(A, \mathbf{P}_{\Sigma A}\mathbf{X})$. Therefore we can and do restrict the lower right corner of the square to the null component. But the component of the null map in $\operatorname{map}_*(A, \mathbf{P}_{\Sigma A}\mathbf{X})$ is contractible. This is true, since this component is connected and its loop $\Omega \operatorname{map}_*(A, \mathbf{P}_{\Sigma A}\mathbf{X}; \text{null})$ is by adjunction just $\operatorname{map}_*(\Sigma A, \mathbf{P}_{\Sigma A}\mathbf{X}) \simeq *$, as needed. Now a pull-back square with two lower corners contractible must have a top arrow that is an equivalence, as needed.

Under the assumption $[A, X] \simeq *$ we get, of course, the same conclusion since the proof above works as well since the right-most arrow in (B.3.2) above is again null.

C. Examples of cellular spaces

Using the adjunction relations (A.2) and Theorem (B.2) one can prove that certain spaces are A-cellular with respect to an appropriate space A. In particular, many well-known constructions in classical homotopy theory lead to A-cellular space.

C.1 JAMES FUNCTOR JX. James, in his thesis [J], gave a combinatorial construction of a space that, under mild conditions, is homotopy equivalent to the loops on the suspension of the given space. In spite of the combinatorial nature of the original definition it is not hard to see both directly and indirectly that:

CLAIM: *For any X the space $JX \simeq \Omega\Sigma X$ is an X-cellular space, in a formula:*

$$\mathbf{CW}_X JX \simeq JX.$$

Proof: First we give an explicit construction of JX as a homotopy colimit of X-cellular spaces [Bl-1, 3.5]. We have a filtration

$$X \subset \cdots \subset J_n X \subset J_{n+1} X \subset \cdots \subset JX$$

We first define an intermediate space $T_n X$ as a homotopy pushout of pointed spaces along the obvious inclusion maps:

$$\begin{array}{ccc} J_{n-1}X & \longrightarrow & J_{n-1}X \times X \\ \downarrow & & \downarrow \\ J_n X & \longrightarrow & T_n X \end{array}$$

and then complete the inductive construction of $J_{n+1}X$ as a homotopy pushout of pointed spaces below, where the maps on $T_n X$ are induced by the corresponding maps from $J_n X$ and $J_{n-1}X \times X$ that agree on $J_{n-1}X$ in the above pushout square; see comments on pushout squares in Appendix HL below:

$$\begin{array}{ccc} T_n X & \longrightarrow & J_n X \times X \\ \downarrow & & \downarrow \\ J_n X & \longrightarrow & J_{n+1} X \end{array}$$

This gives an inductive definition of $J_n X$. Since by Theorem 2.D.16 above a product of two X-cellular spaces is an X-cellular space, we get by induction that $J_{n+1}X$ is an X-cellular space. Therefore $JX = \operatorname{hocolim}_I J_n X$ is also an X-cellular space.

This gives an explicit construction of JX as a pointed hocolim starting with X, i.e. as a member of $\mathcal{C}^{\cdot}(X)$. But the abstract fact that $\Omega\Sigma X \in \mathcal{C}^{\cdot}(X)$ can be obtained directly from (3.B.2) above: Simply compute $\mathbf{CW}_X \Omega\Sigma X \simeq \Omega \mathbf{CW}_{\Sigma X} \Sigma X = \Omega\Sigma X$.

COUNTEREXAMPLE: It goes without saying that if X is connected then ΩX is never an X-cellular, since it has a lower connectivity than X itself (2.D.5).

C.2 HILTON–MILNOR–JAMES DECOMPOSITION. The famous theorem of Milnor and Hilton that followed a similar result by James provides a decomposition of $\Sigma\Omega\Sigma X$ for an arbitrary pointed X as a wedge of smash-powers of ΣX and X itself. Thus it gives an explicit description of $\Sigma\Omega\Sigma X$ as an X-cellular space. (Any smash-power of W is W-cellular (2.D.8).)

Using the adjunction relation (A.2) yields immediately that $\Sigma\Omega\Sigma X$ is in fact a ΣX-cellular without, however, saying anything about the nature of the decomposition. Using (C.1) above one computes:

$$\begin{aligned}\mathbf{CW}_{\Sigma X}\Sigma\Omega\Sigma X &\simeq \overline{W}\mathbf{CW}_X\Omega\Sigma\Omega\Sigma X\\ &\simeq \overline{W}\Omega\Sigma\Omega\Sigma X \simeq \Sigma\Omega\Sigma X.\end{aligned}$$

Here we used (C.1) in the second equivalence, so that applying twice the James functor to X still gives an X-cellular space. In fact, the claim follows also immediately from (C.1) above: given that $\Omega\Sigma X$ is X-cellular it follows immediately that $\Sigma\Omega\Sigma X$ is ΣX-cellular as claimed.

Notice, however, that for non-suspension $\mathbf{CW}_Y\Sigma\Omega Y \not\simeq \Sigma\Omega Y$. In fact $\Sigma\Omega Y$ is not a Y-cellular space, rather the other way around: As we shall see X is $\Sigma\Omega X$-cellular (C.6)(C.7). For example, $\Sigma\Omega K(\mathbb{Z},3) \simeq \Sigma\mathbb{C}P^\infty$ is not $K(\mathbb{Z},3)$-cellular since any $K(\mathbb{Z},3)$-cellular space must have vanishing reduced complex mod-p K-theory [An-H] and $\Sigma\mathbb{C}P^\infty$ is not K-acyclic.

Similarly $\Omega^n S^n X$ is also an X-cellular space.

C.3 THEOREM: *For any X the Dold–Thom functor $SP^\infty X$ is an X-space.*

Remark: In Chapter 4 below a direct presentation of the symmetric products, finite and infinite, will be given as a pointed homotopy colimit of a diagram made up of finite powers of X, so the theorem will be proved more generally with more elementary means. Here we indicate a proof using a technique of Bousfield.

Proof: This follows from a more general observation about arbitrary 'convergent functors' of [B-F], or Γ-spaces of [S]. Let $\Gamma \subset \mathrm{Sets}_*$ be the full subcategory of the objects $n^+ = \{0,\dots,n\}$ with base point $0 \in n^+$, for $n \geq 0$. A Γ-space is a functor $U\colon \Gamma \longrightarrow \mathcal{S}_*$ that assign the point to 0^+. It is special if the canonical product map: $U(n^+) \simeq U(1^+) \times \cdots \times U(1^+)$, of the maps that send all elements except one to $0 \in 1^+$, is an equivalence and very special if the induced monoid on $\pi_0 U(1^+)$ is an abelian group.

Each Γ-space determines a functor $U\colon \mathcal{S}_* \longrightarrow \mathcal{S}_*$ with $UX = \mathrm{diag}(UX_\bullet)$. where $(UX_k)_\bullet$ is the space associated by the Γ-space U to the set of k-simplices X_k of X. Thus every, very special Γ-space $h\colon \Gamma \longrightarrow \mathcal{S}_*$ determines a reduced homology theory $\pi_* hX \equiv h_* X$.

C.4 PROPOSITION: *For any Γ-space U and any $X \in \mathcal{S}_*$ the space UX is an X-cellular space.*

Proof: Almost by definition U can be written as the 'tensor product' of Γ^{op}-spaces with Γ-space [B-4, 6.4], [B-3, 6.1]:

$$UX \simeq \iint X^{\cdot} \wedge U(\cdot)$$

where $\iint$ denotes the homotopy 'coend' (coequalizer) ([Mac] and Appendix HC below) over Γ. Notice that $X^{\cdot}\colon \Gamma^{op} \longrightarrow$ space is

$$\begin{aligned} X^{n^+} &= X \times \cdots \times X, \quad (n+1) \text{ times;} \\ X^{n^+} &= \mathrm{map}_*(n^+, X), \quad \text{this gives a functor;} \\ \Gamma^{op} &\longrightarrow \text{spaces.} \end{aligned}$$

Now since by (2.D.16) X^{n^+} is an X-space and by Lemma 2.D.8 above $X^{n^+} \wedge Y$ is an X-space for any Y, we get that UX is a pointed hocolim of X-spaces and therefore an X-space.

In order to deduce Theorem C.3 above it is enough to show that $SP^\infty X$ is equivalent to UX for some Γ-space $U\colon \Gamma \longrightarrow \mathcal{S}_*$. But [B-4, 6.2] shows that choosing the discrete Γ-space $\tilde{\mathbf{Z}}$ to be $\tilde{\mathbf{Z}}(n^+) = \mathbf{Z} \oplus \cdots \oplus \mathbf{Z}$ n-times, and regarding $\tilde{\mathbf{Z}}$ as a discrete Γ-space, gives $\tilde{\mathbf{Z}}X \simeq SP^\infty X$. Therefore $SP^\infty X$ is an X-space.

By the same token $\Omega^\infty S^\infty X$ is also an X-space since $\Omega^\infty S^\infty X$ can be obtained as a diagonal of a Γ-space.

Further examples of A-cellular spaces can be derived from the following:

C.5 PROPOSITION: *Let $V(n)$ denote a finite p-torsion space of Hopkins-Smith type n, where in particular $V(n)$ is acyclic with respect to the Morava homology theory $K(n)$. There exists an integer $k = k(n)$ such that, for all $i \geq 0$,*

$$K(G, i+k) = \mathbf{CW}_{V(n)} K(G, i+k).$$

In particular, $K(G, i+k)$ is a homotopy colimit of finite $K(n)$-acyclic subcomplexes.

Proof: The point is that one can show that $\mathbf{P}_{\Sigma V(n)} K(G, n+k) \simeq *$ for large k and then use (B.3) above. To show this equivalence one proceeds by induction. Consider $n = 1$. It follows from the usual cofibration sequence that the mod-p homotopy groups of the $\Sigma V(1)$-nullification must be v_1-periodic (i.e. v_1-local) (compare Chapter 8). But we will see (Chapter 4) that the nullification of any

Eilenberg–Mac Lane space is a product of (at most two) such spaces (4.B.4.1). But a product of non-trivial Eilenberg–Mac Lane spaces cannot be periodic with respect to the v_1 map unless the function complex from both the range and the domain are contractible since all composition operations that change dimensions on such a space must vanish. Moreover, a product of two Eilenberg–Mac Lane spaces cannot have non-trivial homotopy groups in infinite number of dimensions. So mod-p homotopies must vanish from the dimension d for which the Adams map $v_1 : M^{d+q}(\mathbb{Z}/p\mathbb{Z}) \longrightarrow M^d(\mathbb{Z}/p\mathbb{Z})$ exists, namely, if p is odd, for $d \geq 3$ [C-N]. This proves that odd primes, $\mathbf{P}_{\Sigma V(1)}K(\mathbb{Z}/p\mathbb{Z}, 3) \simeq *$. This gives the result for $n = 1$. For higher dimensions one proceeds similarly by induction on the construction of $V(n+1)$ out of $V(n)$.

C.6 Classifying spaces. It is not hard to see directly that Milnor's classifying space construction leads to a description of BG, for any group-space G, as a G-cellular space, i.e. $BG \in \mathcal{C}^{\cdot}(G)$. But this fact is a direct corollary of (A.2) above. In fact, one can prove: $\mathbf{P}_{\Sigma G}BG \simeq *$. To this end use (A.1) to get $\mathbf{P}_{\Sigma G}BG = \overline{W}\mathbf{P}_G\Omega BG \simeq \overline{W}\mathbf{P}_G G \simeq \overline{W}\{*\} = \{*\}$. Therefore $BG = \mathbf{CW}_G(BG)$ as needed. In particular, $K(G, n+k)$ is a $K(G,n)$-space for any $k \geq 1$.

Moreover, BG is always a ΣG-space since (using A.2)

$$\mathbf{CW}_{\Sigma G}BG \simeq B\mathbf{CW}_G\Omega BG = B\mathbf{CW}_G G = BG.$$

From this observation we get also

C.7 Corollary: *Any connected space X in $\mathcal{S}_*$, is in $\mathcal{C}^{\cdot}(\Sigma\Omega X)$.*

In fact it is not difficult to write for a group object $G \in \mathcal{S}_*$, the classifying space $\overline{W}G$ is ΣG-cellular. We use the pointed Borel construction of the suspension of the group G. While G has no fixed point with respect to the self-action by left multiplication, the suspension has two fixed points and so we can take the pointed homotopy colimit of the suspension ΣG. It takes a bit of technique to see that we get again the classifying space of G:

$$\overline{W}G = EG \ltimes_G \Sigma G. \quad \text{(see (9.D.3) below).}$$

C.8 Proposition: *For all $n, k \geq 0$, $K(G, n+k)$ is a $K(\mathbb{Z}, n)$-cellular space.*

Proof: Since $K(G, m+1) = BK(G,m)$ it is sufficient to prove that $K(G,n)$ is always $K(\mathbb{Z}, n)$-cellular. For $n = 1$ it is clear, since any connected $X \in \mathcal{S}_*$ is

S^1-cellular. For $n > 1$, the group G is abelian and thus we have a fibration

$$K(F,n) \to K(F',n) \to K(G,n)$$

with F and F' free abelian groups. Therefore, by (2.D.11) above we can assume that G is a free abelian group. But then we can write $K(F,n)$ as a homotopy limit of an increasing sequence of $K(F_\alpha, n)$ for free abelian subgroups F_α. So we are done by a transfinite induction argument using (2.D.17) to deduce that, if $K(F,n)$ is $K(\mathbf{Z},n)$-cellular, then so is $K(F \oplus \mathbf{Z}, n) \simeq K(F,n) \times K(\mathbf{Z},n)$.

The following example shows that the sequence in (B.2) is not always a fibration sequence:

C.9 EXAMPLE. $\mathbf{CW}_{K(\mathbf{Z}/p\mathbf{Z},1)} K(\mathbf{Z}/p^2\mathbf{Z},1) = K(\mathbf{Z}/p\mathbf{Z},1)$.

Proof: Consider the map g: $\mathbf{Z}/p\mathbf{Z} \longrightarrow \mathbf{Z}/p^2\mathbf{Z}$ of abelian groups $1 \longrightarrow p$. This is a generator of $\mathrm{Hom}(\mathbf{Z}/p\mathbf{Z}, Z/p^2\mathbf{Z}) \simeq \mathbf{Z}/p\mathbf{Z}$, and it induces a map on the classifying spaces: Bg: $K(\mathbf{Z}/p\mathbf{Z},1) \longrightarrow K(\mathbf{Z}/p^2\mathbf{Z},1)$. Since the source is clearly a $K(\mathbf{Z}/p\mathbf{Z},1)$-cellular space it is sufficient to show that Bg induces a homotopy equivalence, namely $\mathrm{map}_*(K(\mathbf{Z}/p\mathbf{Z},1), Bg)$, on the pointed function complexes. But the pointed function complex is homotopically discrete with

$$\mathrm{map}_*(K(\mathbf{Z}/p\mathbf{Z},1),\ K(G,1)) = \mathrm{Hom}(\mathbf{Z}/p\mathbf{Z}, G).$$

Therefore the above map Bg gives us the correct $\mathbf{CW}_A$-approximation for $A = K(\mathbf{Z}/p\mathbf{Z},1)$. In view of (1.H.5) we see that the sequence in (B.2) is not always a fibration sequence.

C.10 COROLLARY: *If $A = K(\mathbf{Z}/p^k\mathbf{Z},n)$ and $X = K(\mathbf{Z}/p^\ell\mathbf{Z},n)$, then*

$$\mathbf{CW}_A X = \begin{cases} A & \text{if } k \le \ell, \\ X & \text{if } k \ge \ell. \end{cases}$$

Proof: This is clear using the above together with the fibration theorem (2.D.11). Thus the fibration

$$K(\mathbf{Z}/p^2\mathbf{Z},n) \xrightarrow{\times p} K(\mathbf{Z}/p^2\mathbf{Z},n) \longrightarrow K(\mathbf{Z}/p\mathbf{Z},n) \times K(\mathbf{Z}/p\mathbf{Z}n+1)$$

by (C.8) and (C.9) above presents the space $K(\mathbf{Z}/p\mathbf{Z},n)$ as a $K(\mathbf{Z}/p^2\mathbf{Z},n)$-cellular space.

C.11 EXAMPLE Let $X = M^{n+1}(p^\ell)$ and $A = M^{n+1}(p)$ be two Moore spaces, with $H_n(M^{n+1}(p^\ell), \mathbb{Z}) = \mathbb{Z}/p^\ell\mathbb{Z}$. Then $\mathbf{CW}_A X$ is a fibre in:

$$F \longrightarrow X \longrightarrow K(\mathbb{Z}/p^{\ell-1}\mathbb{Z}, n),$$

while $\Sigma X = \mathbf{CW}_A \Sigma X$.

Proof: Compare [Bl-2, 3.1] To compute the fibre of the composition $X \longrightarrow K(\pi_n X, n) \longrightarrow K(\mathbb{Z}/p^{\ell-1}\mathbb{Z}, n)$ as $\mathbf{CW}_A X$ we consider the pointed function complex of $M^{n+1}(p)$ into the fibration. Since $\mathrm{map}_*(M^{n+1}(p), K(\mathbb{Z}/p^\ell\mathbb{Z}, n)) \simeq \mathbb{Z}/p\mathbb{Z}$ is homotopically discrete by cohomological computation, we first notice that the fibre has the correct function complex from $M^{n+1}(p)$. We then must show that the fibre is a $M^{n+1}(p)$-cellular space. But the fibre is a p-torsion space so it has a Hilton–Eckmann cell decomposition $M^{n+1}(p) \cup CM^{n+2}(H_{n+1}(F), n+1) \cup \cdots$ where all the attaching maps can be taken to be pointed maps. Now we can use Lemma C.12 below and (B.3) above to conclude that the Moore space $M^{n+2}(H_{n+1}(F, \mathbb{Z}))$ is cellular with respect to $M^{n+1}(p) = M^{n+1}(\mathbb{Z}/p\mathbb{Z})$. This gives a direct representation of F as $M^{n+1}(p)$-cellular since clearly

C.12 LEMMA: *For any p-group G the Moore space $M^{n+j}(G, n+j)$ for $j \geq 2$ is an $M^{n+1}(p)$-space.*

Proof: Use Proposition B.3. Notice that

$$\mathbf{P}_{\Sigma M^{n+1}(p)} M^{n+j}(G, n+j) \simeq *$$

since the localization is an n-connected p-torsion space with all maps from $M^{n+1}(p)$ being null, hence this localization is contractible.

C.13 E_*-ACYCLIC SPACES. The fibration (B.2) relating $\mathbf{P}_{\Sigma A}\mathbf{X}$ and $\mathbf{CW}_A X$ can be used to show that certain E_*-acyclic spaces are $V(n)$-cellular, where $V(n)$ are the spaces introduced by [Sm-1],[Mit] (see also [R]). One can use the following observation:

C.14 OBSERVATION: *Let A be a finite complex with $\tilde{E}_* A \cong 0$. Then for all X the space $\mathbf{CW}_A X$ is the direct limit of its finite E_*-acyclic subcomplexes.*

C.14.1 REMARK: By definition there is a construction of the given acyclic space from A by repeated homotopy colimits. Recall that any pointed homotopy colimit of E_*-acyclic spaces is again E_*-acyclic. The point of the observation is, however, that here there is a direct system of finite E_*-acyclic spaces whose homotopy colimit is equivalent to the given space $\mathbf{CW}_A X$.

Proof: Recall the construction of $\mathbf{CW}_A X$. For a finite A the limit ordinal $\lambda(A)$ is the first infinite ordinal w. Therefore in that case

$$\mathbf{CW}_A X = \lim_{i<\infty}(X_1 \hookrightarrow X_2 \hookrightarrow X_i \hookrightarrow),$$

where X_i are all subcomplexes of $\mathbf{CW}_A X$. But now, by induction, we can show that each X_i is the limit of a finite E_*-acyclic subcomplex. Notice that if A is any E_*-acyclic space then so is the half-suspension $\tilde{\Sigma}^n A = S^n \rtimes A = S^n \times A/S^n \times \{*\}$ by a Mayer–Vietoris argument. Now if by induction $X_j = \varinjlim_{\alpha} A(i)$, where $A(i)$ are finite E_*-acyclic, then since X_{j+1} is a pushout along a collection of maps from $\tilde{\Sigma}^n A$, X_{j+1} is again $\varinjlim_{\beta} A_\beta(j+1)$ where $A(j+1)$ are all finite. This completes the proof.

Our principal tool to detect whether an E_*-acyclic complex X is the limit of its finite E_*-acyclic subcomplexes is the following:

C.15 PROPOSITION: *Let A be an E_*-acyclic finite complex. Then X is the limit of its finite acyclic subcomplexes if $\mathbf{P}_{\Sigma A} X \simeq *$.*

Proof: This is immediate from Theorem B.2 and (B.3) and the lemma above.

It is possible to apply the proposition to the spaces $A = V(n)$ of type $n+1$. Thus $V(0)$ is $S^1 \cup_p e^2$, and for every prime p and $n \geq 0$ there exists a finite p-torsion space $V(n)$ of type n. This means $\tilde{K}(m)_* V(n) = 0$ for all $m < n$ and $\tilde{K}(m)_* V(n) \neq 0$ for all $m \geq n$, where $K(n)$ denotes the n-th Morava K-theory (compare discussion in [M-T], [B-4], [R] and [F-S]).

In Chapter 8 we apply this proposition to detect under what conditions an acyclic space can be constructed from elementary (and finite) ones using [B-4, 9.14 and 13.6].

C.16 REMARK: Using similar techniques one can show (compare 8.B): In the following cases every E_*-acyclic space is in $C^{\cdot}(V(n))$ for an appropriate $n \geq 0$:

(1) For all n there exist $m \geq n$ with $K(G, m+j) \in C^{\cdot}V(n)$ for all j, and all p-torsion groups G.

(2) If $\tilde{K}_{\mathbf{C}}\Omega X \simeq 0$ then $X \in C^{\cdot}(V(n))$ for any p-torsion 2-connected X, see (C.5) above.

(3) For every $n \geq 1$ there exists $N \geq n$ so that if X is N-connected, p-torsion and $\tilde{S}(n)_* \Omega^N X = 0$, then $X \in C^{\cdot}(V(n))$ where $\tilde{S}(n)_*$ is the homology theory from [B-4].

We will not give the proofs here. ∎

D. Localization $\mathbf{L}_f$ and cofibrations, fibrations

In this section we will consider some simple cases where $\mathbf{L}_f$ preserves fibrations and cofibrations. Later on, in Chapter 5, we will consider a more general theory where fibrations are 'almost' preserved by $\mathbf{L}_{\Sigma f}$.

In general there is little hope that $\mathbf{L}_f$ will preserve cofibrations since the cofibre of a map between two f-local spaces is very rarely f-local. This happens more often in the stable category, where sometimes homotopy colimits of E_*-local spaces are E_*-local for certain ('smashing') homology theories E_*.

Nevertheless the following is often useful:

D.1 THEOREM: *Let $\mathbf{L}_f$ be the localization with respect to a map $f \in \mathcal{S}_*$. Let $A \xrightarrow{i} X \xrightarrow{j} X \cup CA$ be a cofibration sequence.*

(i) *If $\mathbf{L}_f A \simeq *$ then $\mathbf{L}_f(j)$ is a weak homotopy equivalence.*

(ii) *If $\mathbf{L}_f(i)$ is a weak equivalence then $\mathbf{L}_f(X \cup CA) \simeq *$.*

Proof: This really follows directly from the general formula for the localization of homotopy colimits (1.D.3) above, but it may be worthwhile to give a direct argument in this simple case.

Consider the cofibration $A \hookrightarrow X \longrightarrow X/A$. We have a factorization of the map into a colimit of the diagram $CA \hookleftarrow A \hookrightarrow X$ given by $X/A \longrightarrow \mathbf{L}_f X/\mathbf{L}_f A \longrightarrow \mathbf{L}_f(X/A)$ with the composition being the coaugmentation of $\mathbf{L}_f$.

Now we assume that $\mathbf{L}_f A \longrightarrow \mathbf{L}_f X$ is a homotopy equivalence so that $\mathbf{L}_f X/\mathbf{L}_f A$ is contractible and $X/A \longrightarrow \mathbf{L}_f(X/A)$ is null homotopic. By the idempotency of $\mathbf{L}_f$ we get $\mathbf{L}_f(X/A) \simeq *$, as needed.

Now assume that $\mathbf{L}_f A \simeq *$. Then we have an equivalence $\mathbf{L}_f X \xrightarrow{\simeq} \mathbf{L}_f X/\mathbf{L}_f A$. Therefore the natural map $X/A \longrightarrow \mathbf{L}_f(X/A)$ factors through a new map $X/A \longrightarrow \mathbf{L}_f X$. By universality we get a map $\mathbf{L}_f(X/A) \longrightarrow \mathbf{L}_f X$, which is easily seen to be a homotopy equivalence as required.

D.1.1 FURTHER CASES: The above results do not treat the more difficult case when we assume $\mathbf{L}_f X \simeq \mathbf{L}_f X/A \simeq *$ and try to conclude something about $\mathbf{L}_f A$. As in the analogous case of fibration this can be shown, for nullification with respect to suspension, to be manageable. See (5.B.4.1).

Fibrations: We now formulate a general theorem about preservation of fibration by $\mathbf{L}_f$. Later on we will find special cases of its usefulness. In what follows we denote by $\overline{\mathbf{L}}_f$ and $\overline{\mathbf{P}}_W$ the homotopy fibre of the coaugmentation maps.

D.2 THEOREM: *Let $F \longrightarrow E \xrightarrow{p} X$ be a fibration with connected $F, E, X \in \mathcal{S}_*$. Assume that $\mathbf{L}_{\Sigma f}X \simeq \mathbf{L}_f X$ and that $\mathbf{L}_{\Sigma f}\overline{\mathbf{L}}_{\Sigma f}X$ is homotopically discrete and further that $\mathbf{L}_f\overline{\mathbf{L}}_f E \simeq *$. Then*

$$\mathbf{L}_f F \longrightarrow \mathbf{L}_f E \xrightarrow{\mathbf{L}_f p} \mathbf{L}_f X$$

is also a fibration sequence.

The following are particularly useful special cases:

D.3 COROLLARY: *Consider the case: $\mathbf{L}_f = \mathbf{P}_W$ for any space W.*

(1) *If $\mathbf{P}_{\Sigma W}X = \mathbf{P}_W X$ then $\mathbf{P}_W$ preserves the fibration sequence, i.e. $\mathbf{P}_W F \longrightarrow \mathbf{P}_W E \longrightarrow \mathbf{P}_W X$ is a fibration sequence.*

(2) *If X is W-null then $\mathbf{P}_W$ preserves the fibration sequence.*

Proof: Case (2) is a special case of (1). In case (1) one uses the fact that $\mathbf{P}_W\overline{\mathbf{P}}_W X \simeq *$ for any A, X. (See (1.H.1) and (1.H.2).) Therefore the assumptions of Theorem D.2 are satisfied and the fibration is preserved by $\mathbf{P}_W$.

Proof of D.2: Consider the following commutative diagram of fibrations built by taking homotopy fibres from the lower right square:

$$\begin{array}{ccccccc}
\Omega\overline{\mathbf{L}}_{\Sigma f}X & \longrightarrow & \overline{F} & \longrightarrow & \overline{L}_f E & \longrightarrow & \overline{L}_f X = \overline{L}_{\Sigma f}X \\
\downarrow & & \downarrow & & \downarrow & & \downarrow \\
\Omega X & \longrightarrow & F & \longrightarrow & E & \xrightarrow{p} & X \\
\downarrow & & \downarrow & & \downarrow & & \downarrow \\
\Omega\mathbf{L}_{\Sigma f}X & \longrightarrow & G & \longrightarrow & \mathbf{L}_f E & \xrightarrow{\mathbf{L}_f p} & \mathbf{L}_f X = \mathbf{L}_{\Sigma f}X
\end{array}$$

Looking at the diagram we see that it suffices to show that $F \longrightarrow G$ induces an equivalence $\mathbf{L}_f F \xrightarrow{\simeq} \mathbf{L}_f G$ since G, being the homotopy fibre of a map between two f-local spaces, is f-local (1.A.8(e.3)). We first show that G is connected. Since by elementary fact (1.A.8) (e.11) G is the homotopy fibre of a map between two connected spaces, it is sufficient to show that $\mathbf{L}_f(p)$ induces a surjective map on the fundamental group. Since Σf is a map of simply connected spaces, the map induced by the coaugmentation $\pi_1 X \to \pi_1\mathbf{L}_{\Sigma f}X$ is surjective. Since the map p itself also induces a surjective map on fundamental groups, because the fibre F is connected, we are done by commutativity.

Since G is connected we can apply (1.H.1) to the fibration $\overline{F} \to F \to G$, to show that $\mathbf{L}_f F \cong \mathbf{L}_f G \cong G$ since G is f-local by (e.3). For this we need to prove $\mathbf{L}_f\overline{F} \cong *$. Our assumptions are designed to achieve just this. Turning to the fibration sequence over $\overline{\mathbf{L}}_f E$ we notice that this space is connected by (e.11), since $\mathbf{L}_f\overline{\mathbf{L}}_f E \cong *$ by assumption. Therefore by (1.H.1) again it is sufficient to show

$\mathbf{L}_f\Omega\overline{\mathbf{L}}_{\Sigma f}X \cong *$, but by commutation (A.1) above the latter localization is equivalent to $\Omega\mathbf{L}_{\Sigma f}\overline{\mathbf{L}}_{\Sigma f}X$, which we assume by homotopy discreteness to be contractible.

D.4 REMARK In view of (3.A.1), the condition $\mathbf{L}\Omega B = \Omega\mathbf{L}B$ is equivalent for a connected space B to $\mathbf{L}_{\Sigma f}B \simeq \mathbf{L}_f B$. Notice that this last condition is hardly ever true unless B is already f-local. There is one exception, however. If $f : S^1 \to S^1$ is a self-map of the circle, then for a simply connected space X it is clear that $\mathbf{L}_f X = \mathbf{L}_{\Sigma f}X$. In that case $\mathbf{L}_f$ is the localization at the subring $\mathbb{Z}[\frac{1}{r}]$ if $\deg f = r$. Therefore the classical result that the Sullivan–Quillen rationalization functor $X \to X \otimes \mathbb{Z}[\frac{1}{r}]$ preserves any fibration of simply connected spaces is a special case of the (D.3) or (D.2) above.

E. $\mathbf{CW}_A$ and fibrations

The commutation $\mathbf{CW}_A\Omega X \simeq \Omega\mathbf{CW}_{\Sigma A}X$ given in (A.3) allows one to prove theorems about preservation of fibration by $\mathbf{CW}_A$ under strong assumptions. Here too we will see in Chapter 5 that by allowing 'error terms' or 'near preservation' one can prove much stronger results with only mild assumptions on the given fibration and A. We will need the following:

E.1 LEMMA: *In any fibration sequence $F \longrightarrow E \longrightarrow X$ in $\mathcal{S}_*$, if X and E are in $\mathcal{C}^{\cdot}(\Sigma A)$ then $F \in \mathcal{C}^{\cdot}(A)$. If F and X are in $\mathcal{C}^{\cdot}(\Sigma A)$ then $E \in \mathcal{C}^{\cdot}(A)$.*

Proof: This is immediate from (2.D.12) and (A.2) above by backing up the fibration.

E.2 THEOREM : *Given any fibration of a pointed space $F \longrightarrow E \longrightarrow B$ one can map the following fibration into it:*

$$\begin{array}{ccccc} \mathbf{CW}_A F & \longrightarrow & \overline{E} & \longrightarrow & \mathbf{CW}_{\Sigma A} B \\ \downarrow{\scriptstyle f} & & \downarrow{\scriptstyle g} & & \downarrow{\scriptstyle h} \\ F & \longrightarrow & E & \longrightarrow & B \end{array}$$

where $\overline{E}$ is A-cellular and g a ΣA-equivalence.

E.3 REMARK: Thus although $\overline{E}$ is A-cellular it is slightly removed from being $\mathbf{CW}_A E$, since it is only ΣA-equivalent to E not A-equivalent to it. We shall see in Chapter 5 that the fibre of a canonical map $\overline{E} \longrightarrow \mathbf{CW}_A E$ associated to such a fibration is a GEM for $A = \Sigma A'$, a suspension space.

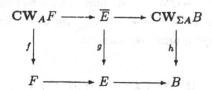

Proof: The proof is easy for principal fibrations with the fibre over the base point being the group $F = G$. One first pulls back the fibration over $\mathbf{CW}_{\Sigma A}B$ to get a principal fibration with the same fibre $G = F$ over the needed base space. This last fibration is classified by a map $\mathbf{CW}_{\Sigma A}B \to \overline{W}G$ to the classifying space of G. But this last classifying map can be factored uniquely up to homotopy through $\mathbf{CW}_{\Sigma A}\overline{W}G \to \overline{W}G$ by the universal properties of $\mathbf{CW}_{\Sigma A}$. The domain of this last map is by (A.2), the classifying space of $\mathbf{CW}_A G$. Therefore we have obtained a fibration over $\mathbf{CW}_{\Sigma A}B$ with the fibre $\mathbf{CW}_A G$ that maps naturally up to homotopy to the given fibration.

The general case is similar, but we need to use the usual technique of expressing the base of a principal fibration as a realization of a bar construction associated with it [B-4],[B-3, 5.5]. We give only the outline.

Here we also use both the equivalence $\mathbf{CW}_A \Omega X \simeq \Omega \mathbf{CW}_{\Sigma A} X$ and the natural equivalence (2.E.8):

$$\mathbf{CW}_A(X \times Y) \simeq \mathbf{CW}_A X \times \mathbf{CW}_A Y.$$

We consider first the associated principal fibration:

$$\Omega B \longrightarrow F \longrightarrow E$$

where we consider E as the 'quotient space' up to homotopy of F under the action $\Omega B \times F \longrightarrow F$. This action gives rise to a map

$$\mathbf{CW}_A(\Omega B) \times \mathbf{CW}_A F \longrightarrow \mathbf{CW}_A F,$$

or $\Omega \mathbf{CW}_{\Sigma A} B \times \mathbf{CW}_A F \longrightarrow \mathbf{CW}_A F$. We now use the equivalence as usual to show that the original action $\Omega B \times F \longrightarrow F$ gives rise to an action of $\Omega \mathbf{CW}_{\Sigma A} B$ on $\mathbf{CW}_A F$. Therefore we get a principal fibration

$$\Omega \mathbf{CW}_{\Sigma A} B \longrightarrow \mathbf{CW}_A F \longrightarrow \overline{E}$$

associated to that action. Classifying this fibration gives us the desired sequence. Since in that last fibration both fibre and total spaces are A-cellular, so is $\overline{E}$ by (2.D.11). Further, since in the map of fibrations Ωh is an A-equivalence and f is an A-equivalence, we get by the usual long exact sequence for A-homotopy groups $[\Sigma^\ell A, -]_*$ that g is an ΣA-equivalence. This completes the proof.

E.4 COROLLARY: *Let $F \longrightarrow E \xrightarrow{p} B$ be any fibration sequence in $\mathcal{S}_*$ and let $\Sigma^2 A$ be any double suspension in $\mathcal{S}_*$. If B is $\Sigma^2 A$-cellular and if p induces the trivial map on homotopy classes from ΣA, namely $[\Sigma A, p] \sim *$, then*

$$\mathbf{CW}_{\Sigma A}F \longrightarrow \mathbf{CW}_{\Sigma A}E \longrightarrow B = \mathbf{CW}_{\Sigma A}B = \mathbf{CW}_{\Sigma^2 A}B$$

is also a fibration sequence.

E.5 EXAMPLE: Take $A = S^1$. Then (E.4) states the easily checked fact that, over a 2-connected space B, one can define a *fibrewise universal covering* space.

Proof of E.4: We consider the diagram:

$$\begin{array}{ccccc}
Y & \xrightarrow{=} & Y & & \\
\downarrow & & \downarrow & & \\
\mathbf{CW}_{\Sigma A}F & \longrightarrow & \overline{E} & \longrightarrow & \mathbf{CW}_{\Sigma^2 A}B = \mathbf{CW}_{\Sigma A}B \\
{\scriptstyle f}\downarrow & & {\scriptstyle g}\downarrow & & \downarrow \\
F & \longrightarrow & E & \longrightarrow & B
\end{array}$$

This diagram is constructed using the theorem above. We claim that there is a natural equivalence: $\bar{e}\colon \overline{E} \longrightarrow \mathbf{CW}_{\Sigma A}E$. First notice that since B is $\Sigma^2 A$-cellular we get from Lemma E.1 that $\overline{E}$ is ΣA-cellular. Therefore there is a unique natural factorization $\bar{e}$. To complete the proof one must check that $\bar{e}$ is a homotopy equivalence. To this end we use the A-Whitehead theorem and check that the map

$$\operatorname{map}_*(\Sigma A, \bar{e})\colon \operatorname{map}_*(\Sigma A, \overline{E}) \longrightarrow \operatorname{map}_*(\Sigma A, \mathbf{CW}_A E)$$

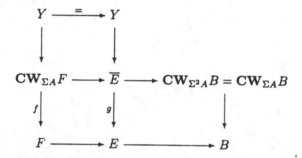

is a homotopy equivalence. Notice that $\operatorname{map}_*(\Sigma A, Y) \simeq *$ since Y is the fibre of f and $\operatorname{map}_*(\Sigma A, f)$ is a homotopy equivalence. We know from the universal property that f is a ΣA-equivalence. Thus $\operatorname{map}_*(\Sigma A, Y) \simeq *$ since Y is the fibre of a ΣA-equivalence f. Now this shows that the homotopy fibre of the map $\operatorname{map}_*(\Sigma A, \bar{e})$ is contractible. This is not enough since the base space $\operatorname{map}_*(\Sigma A, \mathbf{CW}_A E)$ is not connected. To complete the proof one must show that the induced map of function complexes on the level of components, i.e. on π_0, is an isomorphism. Since all the components of function complexes out of suspensions are equivalent to each other. Our assumption guarantees that it is surjective on components since it implies by a simple diagram chase that any map from ΣA to E lifts up to homotopy to a map $\Sigma A \longrightarrow \overline{E}$. Injectivity follows from the above-mentioned property of Y.

4. DOLD–THOM SYMMETRIC PRODUCTS AND OTHER COLIMITS

Introduction

One of the main results of this chapter says that if X is an infinite loop space that is equivalent (as such) to a possibly infinite (weak) product of $K(\Pi, n)$'s with Π an abelian group and $n \geq 0$ (here called GEM), then so are both $\mathbf{L}_f X$ and $\mathbf{CW}_A X$ for any $f\colon A \longrightarrow B$ and $A \in \mathcal{S}_*$. Moreover, these functorial constructions inherit the abelian group structure that X possesses as a GEM and the coaugmentation and augmentation maps are essentially GEM maps, namely up to homotopy they can be realized as homomorphisms in the category of simplicial abelian groups. This implies immediately that when applying $\mathbf{L}_f$ or $\mathbf{CW}_A$ to an Eilenberg–Mac Lane space $K(G, n)$ with G an abelian group, the resulting GEM has at most two non-vanishing homotopy groups and they are in adjacent dimensions. These results would follow immediately if we would have a version of functors $\mathbf{L}_f$ and $\mathbf{CW}_A$ that strictly commutes with products since then the functors would turn (simplicial) abelian groups into such and the (co)augmentation map would be a group map. So part of the following development is there because the present version of these functors commutes with finite products only up to homotopy. This property of idempotent functors is a step towards a better understanding of the behaviour of fibration sequences under $\mathbf{L}_f$, $\mathbf{CW}_A$ and, in particular, it will lead to an understanding of the behaviour of fibrations under Bousfield's homological localizations for any generalized homology theory. The above result is proved by presenting the symmetric product on X as a homotopy colimit of a certain diagram consisting of finite powers of X even though they are defined as strict direct limits (quotient spaces) of some finite power of X. The question then arises about the general relation between the colimit and the homotopy colimit of diagrams of (pointed) spaces.

This leads to the second concern in this chapter: We will 'estimate' the difference between the homotopy colimit (pointed) and the strict colimit of pointed diagrams in $\mathcal{S}_*$. We do that by estimating the cofibre of the natural map $\operatorname{hocolim}_* \underset{\sim}{X} \longrightarrow \operatorname{colim} \underset{\sim}{X}$.

By an 'estimate' we mean a cellular inequality of the type $X \ll SP^k X$ (2.D.2.1) or

$$\Sigma^2 A \ll \operatorname{cofibre}(\operatorname{hocolim}_* \underset{\sim}{X} \longrightarrow \operatorname{colim} \underset{\sim}{X}).$$

Namely, the symmetric products on X are X-cellular and the above cofibre is $\Sigma^2 A$-cellular.

This inequality immediately implies, for example, that the said cofibre is always 1-connected because double suspensions are always 1-connected and cellular equalities preserve connectivity.

A. Dold–Thom symmetric products as homotopy colimits

Given an A-cellular space X equipped with an action by a group G, we may ask under what condition the quotient space $X/G = \operatorname{colim}_G X$ is also A-cellular.

Before entering the discussion let us recall the concept of cellular inequality, which is clearly a partial order on spaces.

A.1 RECALL (2.D.2.1): We say that X is built from A and write $A \ll X$ if X is A-cellular.

A more extensive discussion of $\ll$ is given in Chapter 8 below.

Our main result is not unexpected: We start with a topological or simplicial group G —for our purpose it is sufficient to consider a discrete, finite group— and a G-space X that has at least one G-fixed point. We assume that the action preserves the cell structure of the space X.

A.2 THEOREM: *Let X be a pointed G-space. Assume that for all $H \subseteq G$ the sub-space X^H of H-fixed points in X is A-cellular. Then so is the quotient space X/G.*

Remark: Thus in the above notation the theorem says that if, for all $H \subseteq G$, $A \ll X^H$ then $A \ll X/G$. In particular we can apply this to symmetric products. In this partial order notation we now deduce that for all X one has the inequality $X \ll SP^k X$. In the following we denote by S_k the symmetric group on k letters.

A.2.1 COROLLARY: *Let X be any pointed space. For all $0 \le k \le \infty$ the k-fold symmetric product $SP^k X = X^k/S_k$ is X-cellular space.*

A.2.2 COROLLARY: *If X is an E_*-acyclic space where E_* is any homology theory then so is the k-fold symmetric product on X for all $0 \le k \le \infty$.*

Proof of Corollaries A.2.1 and A.2.2: The second follows immediately from examples (3.D.2.1) and (3.D.2.5). As for the first, by the theorem above one need only check that for any subgroup $H \subseteq \Sigma_k$ of the permutation group Σ_k, the space $(X^k)^H$ is X-cellular. But since H operates on the factors of X^k, it either leaves a whole factor in its place or moves it wholly to another copy of X. Thus it is clear that all the fixed point subspaces are also product $X \times \cdots \times X$ (ℓ-times) for some $\ell \le k$. But we saw in (2.D.16) above that X^ℓ is X-cellular for any X. The corollary follows.

A.2.3 EXAMPLE: Thus as defined the space $SP^2 X$ is a quotient of $X \times X$ under the action of Σ_2, the permutation group on two letters. In other words, it is the strict colimit of a diagram $\{X \times X; \tau\}$ with one space and one non-identity self-map. But the argument proving (A.2) given below presents the same symmetric product, up to homotopy equivalence, as a homotopy colimit of a diagram consisting of two spaces:

$$\{X \longrightarrow X \times X;\ \delta, \tau\}$$

and two non-identity maps, one the diagonal δ from X to $X \times X$ the other being the above self-map τ of $X \times X$ permuting the two factors. The essential difference between these two diagrams is that the second one is *free* as a diagram (Appendix HL) while the first one is not. These two diagrams have the same strict colimit, SP^2X, but only in the latter is this colimit equivalent to the homotopy colimit of the same diagram.

Proof of Theorem A.2: One shows that there is a pointed diagram of A-cellular spaces whose pointed homotopy colimit is X/G. The diagram is assembled from the subspaces of fixed points of all the subgroups of G ([DF-1], [D-K-2] and Appendix HL below), namely one first considers the category $O = O(G)$ of G orbits. This is simply the small category of all the quotients G/H for $H \subseteq G$ taken as G-sets with the natural left action of G. The morphism sets in that category O are the G-maps of these G-sets. We get the usual diagram of fixed point spaces $\{X^H\}_{H\in O} \overset{\text{def}}{=} X^O$. Since X has a G-fixed point, this point is common to all the subspaces X^H, thus the diagram is pointed, i.e. it is a functor $X^O : O^{op} \longrightarrow S_*$ from the opposite category of orbits O^{op} to pointed spaces. Notice the canonical equality $X^H = \hom_G(G/H, X)$. We now observe that $\operatorname{colim}_{O^{op}} X^O$, the direct limit of the diagram of fixed points, is in fact isomorphic to X/G, since any two points identified by an action of a subgroup are certainly identified by the action of G itself.

On the other hand, we claim that for the diagram X^O the natural map

$$\operatorname*{hocolim_*}_{O^{op}} X^O \longrightarrow \operatorname*{colim}_{O^{op}} X^O$$

is a homotopy equivalence.

This is the content of Lemma A.3 below. Given that lemma, our proof is finished since we assume that the diagram X^O over the orbit category of G consists of pointed A-cellular spaces, therefore by definition the pointed homotopy colimit $\operatorname*{hocolim_*}_{O(G)^{op}} X^O$ is also an A-cellular space and, by the lemma below, this homotopy colimit is homotopy equivalent to X/G.

The following lemma compares *pointed* and *unpointed* hocolim with strict colim:

A.3 LEMMA: *Let X be a pointed G-space. Then the natural maps*

$$\operatorname*{colim}_{O(G)^{op}} X^O \longleftarrow \operatorname*{hocolim}_{O(G)^{op}} X^O \longrightarrow \operatorname*{hocolim_*}_{O(G)^{op}} X^O$$

are homotopy equivalences.

Proof: As for the map on the left from $\operatorname*{hocolim}_I X^O$ to $\operatorname{colim} X^O$, it is a homotopy equivalence since the diagram of fixed points is a *free* $O(G)^{op}$-diagram of spaces: In general, for any free diagram Y of (cofibrant) spaces over any small category I the canonical map from the homotopy colimit to the colimit itself is a weak homotopy equivalence. Compare [DF-1], [D-K-2], and Appendix HL.

The map on the right is a homotopy equivalence because the nerve of $O(G)$ is contractible, since $O(G)$ has a terminal object G/G, the one-point orbit. Therefore by (2.D.3) above and Appendix HL, the natural map from the unpointed hocolim to the pointed one is a homotopy equivalence.

We now use the above argument to make a direct comparison between the *pointed* Borel construction $EG \ltimes_G X$ and the quotient space X/G.

A.4 The pointed Borel construction: One way to gain a better understanding of quotient spaces, such as symmetric products of X, is to consider the difference between the quotient space and the homotopy quotient space. The latter is just a special case of homotopy colimit: We just view the G-space X as a pointed diagram over the usual small category with one object defined by G. Notice that if $*$ is a base point in X, then the point $(*, *, \ldots, *)$ is fixed by the action of any permutation group on the product $X \times X \times \cdots \times X$. Therefore we are in a situation to take pointed homotopy colimits and we can apply it to spaces that have a G-fixed point. The pointed Borel construction (see Example 3.C.7 above) is just a special case of a pointed homotopy limit. We define $EG \ltimes_G X$, the pointed Borel construction on the pointed G-space X, simply as the mapping cone of the natural cofibration: $BG = EG \times_G \{*\} \longrightarrow EG \times_G X$. Equivalently, it is $(EG \ltimes X)/G$.

A.4.1 EXAMPLE: For any group object $G \in \mathcal{S}_*$, if one considers the left action of G on itself, it induces an action of G on the suspension ΣG, which makes the latter into a pointed G-space. As we shall see below (9.D.3). the associated pointed Borel construction $EG \ltimes_G \Sigma G$ is naturally homotopy equivalent to BG.

Our main result in this section is:

A.5 THEOREM: *Let X be a pointed G-space and let $A \in \mathcal{S}_*$. Assume that for all $H \subseteq G$ the fixed-point subspace X^H is A-cellular. Then the homotopy cofibre C of the natural map*

$$EG \ltimes_G X \longrightarrow X/G \longrightarrow C$$

is $\Sigma^2 A$-cellular.

A central example for the above concerns the action of the symmetric group S_n on the power space $X^n = X \times X \times \cdots \times X$ (n-times). Here we consider X^n as an S_n-diagram and, if X has a base point $(*)$, then $(*, *, \cdots, *)$ is fixed under S_n, thus X^n is a pointed diagram. Since $\mathrm{map}_{S_n}(e, X^n) = X^k$ for some $k \geq 0$, we have $X \ll \mathrm{map}_{S_n}(e, X^n) = (X^n)^e$ for all the orbits $e = S_n/H$ with $H \subset S_n$ of S_n.

Therefore, this special case says:

A.5.1 COROLLARY : *Let X be a connected space in S_*. Then*

$$\Sigma^2 X \ll \mathrm{cof}(ES_n \ltimes_{S_n} X^n \longrightarrow X^n/S_n)$$

where $X^n/S_n = SP^n X$ is the Dold–Thom symmetric product of X.

Remark: This result is just a cellular version of a similar result of Dwyer [Dw-2]. It is not hard to see that under our assumption this natural map induces an isomorphism on the fundamental group, and surjection on π_2. In fact the homotopy fibre is 1-connected if X^H is connected.

Proof: The proof starts with Lemma A.3 above. We want to write both the homotopy colimit, i.e. the pointed Borel construction, and the actual orbit space X/G as the same kind of homotopy colimits. This allows one to write the cofibre also as a pointed homotopy colimit and it will be apparent from the latter presentation that we get a $\Sigma^2 A$-cellular space as needed.

The kind of pointed homotopy colimits we need here is so-called *coend*, or rather homotopy coend (Appendix HC, [H-V]). This is the basic construction that allows one to actually write down any G-space as a homotopy colimit of a diagram that depends on the collection of fixed-point subspaces as a diagram over the orbit category of G.

In order to write the pointed Borel construction as well as X/G as a pointed homotopy colimit (more accurately as a homotopy coend), we start by writing the G-space X itself as such as colimit.

For this we recall a basic construction from the homotopy theory of G-spaces [DF-1] (see also Appendix HC). We can assume that X is a $G - CW$ space in the usual sense and so X, being a G-space, can be reconstructed from the diagram of fixed points $\{X^H\} = O^{op}$ with $O^{op} = O(G)^{op}$ the opposite category of G-orbits. Recall that $X^H = \mathrm{map}_G(G/H, X)$, by definition, so X^H is the 'space of all orbits in X of type G/H' if we multiply—half-smash—this space by G/K, which we take now as G-space (unpointed); we get the *pointed* G-space, $G/H \ltimes X^H$. Now whenever one has a diagram $\underset{\sim}{W}$ over a category I and a pointed diagram $\underset{\sim}{V}$ over I^{op}, the opposite category, one can form the coend by gluing together half smashes $\underset{\sim}{W}(i) \ltimes \underset{\sim}{V}(j) = \underset{\sim}{W}^+(i) \wedge \underset{\sim}{V}(j)$. (See Appendix HC.)

NOTATION: We denote the pointed (homotopy) coend of $\underset{\sim}{W}$ and $\underset{\sim}{V}$ by

$$\iint_I \underset{\sim}{W} \ltimes \underset{\sim}{V}.$$

This double integral should remind the reader that there are two diagrams involved in its formation.

Notice (Appendix HC) also that one can write a general homotopy colimit as a homotopy coend by

$$\int_I \underset{\sim}{V} = \iint_{I^{op}} * \ltimes \underset{\sim}{V}.$$

In our case we take the coend of the diagram of all the orbits $G/H_{H \subseteq G}$ thought of now as a diagram of G-spaces over the orbit category O. We denote this diagram by $\mathcal{I}_O$ since it is, by inclusion, a functor from G-orbits to G-spaces, assigning to every orbit thought of as an object in O the orbit itself thought of as a G-space. Denote by X^O the diagram over O^{op} consisting of the fixed-point spaces.

We can sum up the above discussion by:

A.5.2 PROPOSITION: *There is a natural G-equivalence*

$$\iint_O \mathcal{I}_O \ltimes X^O \longrightarrow X$$

from the coend to X.

Similarly one can express X/G as a coend as follows:

A.5.3 PROPOSITION: *There is a natural G-equivalence*

$$\iint_O * \ltimes X^O \xrightarrow{ev} X/G$$

from the coend to X/G.

Proof: This is clear from (A.3) above.

Now we can use this to express the pointed Borel construction as follows:

We need the diagram of classifying spaces $\{BH\}_{H\subseteq G}$. This diagram can be viewed as an O-diagram, since there is a canonical homotopy equivalence $BH \simeq EG \ltimes_G G/H$.

We denote this diagram by {Bh}.

A.6 CLAIM: *There is a homotopy equivalence*

$$EG \ltimes_G X \simeq \iint_O \{Bh\} \ltimes X^H$$

where by $\{Bh\}$ *we mean the O-diagram of the classifying spaces* $\{BH\}$ *as above.*

Proof: One simply substitutes the expression for X as a coend of the fixed-point sets in the pointed Borel construction and use the commutation of homotopy colimits:

$$EG \ltimes_G X \simeq EG \ltimes_G \iint_O \mathcal{I}_O \ltimes X^O$$
$$\simeq \iint_O (EG \times_G G/K) \ltimes X^H \simeq \iint_O \{Bh\} \ltimes X^H,$$

where in the third term of the equation the values of the diagrams on typical orbits G/K and G/H are displayed. We have used the usual expression for BK, the classifying space on K, as the (unpointed) Borel construction on G/K. We have also used the natural equivalence

$$A \ltimes (B \ltimes X) \simeq (A \times B) \ltimes X,$$

for a pointed X and unpointed A, B.

This is a special case of the commutation rule for pointed and unpointed hocolim (2.D.6):

$$(A \times B) \ltimes X \simeq (\oint_A B) \ltimes X \simeq \int_A (B \ltimes X) \simeq A \ltimes (B \ltimes X).$$

To continue the proof:

CLAIM: There is a natural cofibration (as follows) associated with every pointed G-space given in terms of homotopy coends, see Appendix HC below:

$$\text{(A.6.1)} \qquad \iint_{O} \{Bh\} \ltimes X^{O} \longrightarrow \iint_{O} pt \ltimes X^{O} \longrightarrow \iint_{O} \{Bh\} * X^{O}$$

which is naturally equivalent term for term to the natural sequence in the theorem, so that the homotopy coend on the right is equivalent to C from (A.5).

Proof of Claim: By claim A.6 above the first homotopy coend term in (A.6.1) on the left is $EG \ltimes_G X$. The natural map $BH \longrightarrow pt$ then induces by Lemma A.5.2 above the map $EG \ltimes_G X \longrightarrow X/G$, since this is precisely the map that sends every orbit in X to a point. To continue with the proof of the claim we now use the

A.7 OBSERVATION: *For any pointed T and an unpointed S there is a homotopy cofibration sequence:*

$$S \ltimes T \longrightarrow T \longrightarrow S * T$$

where $S * T$ denotes here the reduced join, i.e., it is obtained from the join by collapsing a copy of the cone over S in the usual join to a point, and thus is equivalent to the usual join.

Proof: This follows directly from the simple observation that the mapping cone of the projection $S \times T \longrightarrow T$ is $T * S \vee \Sigma S$. This is clear from the following: If S is unpointed space and T a pointed one, one gets an immediate cofibration square built from the upper left corner square by taking cofibres:

$$\begin{array}{ccccc} S & \longrightarrow & * & \longrightarrow & \Sigma S \\ \downarrow & & \downarrow & & \downarrow \\ S \times T & \longrightarrow & T & \longrightarrow & S * T \vee \Sigma S \\ \downarrow & & \downarrow{=} & & \downarrow \\ S \ltimes T & \longrightarrow & T & \longrightarrow & S * T \end{array}$$

where the space $T * T$ is best thought of as the reduced join, and is pointed. This completes the proof of the observation.

Therefore one has a homotopy cofibration sequence of pointed spaces:

$$BH \ltimes X^{K} \longrightarrow X^{K} \longrightarrow BH * X^{K}$$

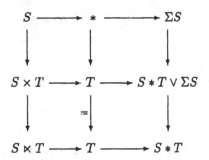

for every two orbits $G/H, G/K \in O = O(G)$. But taking pointed homotopy colimits – and pointed homotopy coends in particular – commutes with taking mapping cones, which are, themselves homotopy colimit. Thus we get the desired cofibration above by taking the homotopy coend over O.

This concludes the identification of the coend on the right in the cofibration (A.6.1) with C in the theorem. This presents C as a pointed homotopy coend and thus as a pointed homotopy colimit of a diagram consisting of the joins. To conclude the proof of the theorem it is sufficient to show that for all H the join space $BH * X^H \simeq S^1 \wedge BH \wedge X^H$ is $\Sigma^2 A$-cellular, since by definition any pointed homotopy colimit on $\Sigma^2 A$-cellular spaces is $\Sigma^2 A$-cellular. But $S^2 = S^1 \wedge S^1 \ll S^1 \wedge BH$ since BH is connected. By assumption one has $A \ll X^H$ for all $H \subseteq G$. Therefore one can conclude by (2.D.8) that

$$\Sigma^2 A \simeq S^2 \wedge A \ll S^2 \wedge X^H \ll S^1 \wedge BH \wedge X^H \simeq BH * X^H,$$

as needed. This completes the proof. ∎

B. Localization and cellularization of GEMs

In this section we will start to consider the relations between generalized Eilenberg–Mac Lane space localization and cellular approximations. To facilitate the discussion we consider here pointed, connected spaces since on them one has a better control on the infinite symmetric product. We prove that both $\mathbf{L}_f$ and $\mathbf{CW}_A$ turns a GEM into a GEM. In fact, the coaugmentation and augmentation maps can be thought of, up to homotopy, as group maps in the category of simplicial abelian groups. Furthermore we will see that, when applied to an Eilenberg–Mac Lane space $K(\pi, n)$ with π abelian, both these functors produce a product of at most two Eilenberg–Mac Lane spaces in adjacent dimensions.

B.1 DEFINITION: Compare [S-V, 2.2 (iii)]. *By a GEM space we mean an infinite loop space that is a generalized Eilenberg–Mac Lane space as a loop space, i.e. it is equivalent as an infinite loop space to a (possibly infinite, weak) product of $K(G,n)$'s taken as abelian group spaces for discrete abelian groups G and for $n \geq 0$. A map $\mathbb{G}_1 \longrightarrow \mathbb{G}_2$ between two GEM spaces is called a GEM map if it is equivalent to a group homomorphism between the corresponding abelian group spaces.*

B.1.1 REMARK: A homotopy characterization of these structures appears in [S-V]. We see from the definition that the homotopy category of GEM spaces and GEM maps is equivalent to the homotopy category of simplicial abelian groups with group maps as morphisms. The latter is in turn equivalent to the homotopy category of chain complexes of abelian groups [May-1]. Thus, for example, if there is a non-null GEM map $K(G, i) \longrightarrow K(G', j)$ then j is either i or $i+1$.

Our aim first is to show that $\mathbf{L}_f(\text{GEM}) = \text{GEM}$ and $\mathbf{CW}_A(\text{GEM}) = \text{GEM}$, i.e. the functors $\mathbf{L}_f$, $\mathbf{CW}_A$ turn GEM spaces into (generally different) GEM spaces. (Compare [B-K, VI.2.2] and [B-5].)

Our main tool in analyzing the effect of $\mathbf{L}_f$ and $\mathbf{CW}_A$ on a GEM is Dold–Thom symmetric products [D-Th], [B-K]. Recall that the infinite symmetric product $SP^\infty X$ on a pointed connected space X is just the direct limit of $SP^k X = X^k/S_k$. This direct limit is clearly also a pointed homotopy colimit. Thus

$$SP^\infty X = \operatorname*{hocolim}_{k<\infty}{}_* SP^k X.$$

It is well known that $SP^\infty X$ is a GEM. In fact it is homotopy equivalent to the infinite product $\prod_{0<i<\infty} K(H_i(X, \mathbb{Z}), i)$. Moreover, simplicially it can be built as the free abelian (simplicial) group on the simplicial set X which we denote here as $\mathbb{Z}X$. Thus we have a coaugmented functor $X \to SP^\infty X$, which is a stepping stone for many of the constructions and theorems of [B-K]. This functor is in a weak sense the universal GEM to which X is mapped: It is easy to see that up to homotopy any map of any space X to a GEM space $X \to \text{GEM}$ factors (non-uniquely) up to homotopy through $X \to SP^\infty X$. Further, it is well known that $SP^\infty X$ has the following universal property:

B.2 Theorem (GEMs as retracts): *If X is a retract up to homotopy of a GEM space $\mathbb{G}$ then X is the underlying space of a GEM. Furthermore, any map $X \longrightarrow \mathbb{F}$ in $\mathcal{S}$ of X to a GEM space $\mathbb{F}$ factors up to homotopy through the natural coaugmentation map $X \longrightarrow SP^\infty X$ of X to the canonical GEM space associated with X.*

Proof: The theorem reduces, as we shall see, to the following lemma.

B.2.1 Lemma *A space X is the underlying space of a GEM if and only if the natural map $X \longrightarrow SP^\infty X$ has a left homotopy inverse.*

Proof of lemma: (Arguing simplicially) If X is the underlying space of a GEM there exist a simplicial abelian group $\mathbb{A}$ that is equivalent to X in $\mathcal{SS}$. But being an abelian group there is an obvious (strict) left inverse to $\mathbb{A} \longrightarrow \mathbb{Z}\mathbb{A}$. Hence we have a homotopy inverse for X. Now suppose the above map has a homotopy inverse. In that case the homotopy groups inject into the integral homology groups since [May-1], [D-Th], [B-K] $\pi_i \mathbb{Z}X \cong H_i(X, \mathbb{Z})$ is a natural isomorphism. One deduces easily using the universal coefficient theorem for cohomology with coefficients in the homotopy groups of X that there is a weak equivalence $X \xrightarrow{\simeq} \prod_{0\leq n<\infty} K(\pi_n X, n)$

by constructing maps to the individual factors. (Compare [Mo, 3.29].) This follows from the fact that given an element $u \in H^k(X, G)$ represented by a map $X \longrightarrow K(G, k)$ the induced map on the k-th homotopy groups factors through the induced map on homology via the Hurewicz map and the image of the element u in the group $Hom(H_k(X, \mathbb{Z}), G)$. ∎

Using this Lemma we now give:

Proof of B.2: If X is a retract of a GEM $\mathbb{G}$ we can construct the following ladder using (B.2.1):

$$
\begin{array}{ccccc}
X & \longrightarrow & \mathbb{G} & \longrightarrow & X \\
\downarrow & & \uparrow\downarrow & & \downarrow \\
\mathbb{Z}X & \longrightarrow & \mathbb{Z}\mathbb{G} & \longrightarrow & \mathbb{Z}X.
\end{array}
$$

Where the retraction in the middle arrow is guaranteed by the lemma and it gives the a section of $X \longrightarrow \mathbb{Z}X$. Thus, again by the lemma above, X is an underlying space of a GEM. The second claim of the theorem is an immediate consequence.

The two main theorems of this section are

B.3 Theorem: *Let $X, A \in \mathcal{S}_*$. Assume X is a GEM. Then $\mathbf{CW}_A X$ is also a GEM. If, in addition, $\pi_i X \simeq 0$ for $i \geq N$, then $\pi_i \mathbf{CW}_A X \simeq 0$ for $i \geq N$. Moreover, the augmentation map $j : \mathbf{CW}_A X \to X$ is, up to homotopy, a retract of $SP^\infty(j)$. Furthermore, the same holds for the homotopy idempotent functor $\overline{\mathbf{P}}_A$, i.e. the homotopy fibre of nullification (2.A.3.1) (2) above.*

B.4 Theorem: *Let $f \colon A \longrightarrow B$, $X \in \mathcal{S}_*$. Assume X is a GEM. Then $\mathbf{L}_f X$ is also a GEM. If, in addition, $\pi_i X \simeq 0$ for $i \leq N$, then $\pi_i \mathbf{L}_f X \simeq 0$ for $i \leq N$. Moreover, the coaugmentation map $j : X \to \mathbf{L}_f X$ is, up to homotopy, a retract of $SP^\infty(j)$.*

B.4.1 Corollary: *For any $K(\pi, n)$ with π abelian group, the homotopy groups of $\mathbf{CW}_A K(\pi, n)$ vanish in all dimensions except possibly $n-1$ and n; the homotopy groups of $\mathbf{L}_f K(\pi, n))$ vanish in all dimensions except possibly n and $n+1$.*

Proof of corollary: (Compare [B-K, VI.2.2] This follows at once from Remark B.1.1 above and (1.C.4) since the projection of the (co)augmentation map to any other Eilenberg–Mac Lane factors must be null homotopic. ∎

Both proofs rely heavily on (4.A.2.1) above, namely on $X \ll SP^k X$ for all $0 \leq k \leq \infty$. In both (B.3) and (B.4) the vanishing of π_i follows easily from universality.

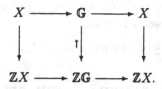

Proof of B.3: Let $X \longrightarrow SP^\infty X \overset{r}{\dashrightarrow} X$ be the inclusion followed by the homotopy retraction guaranteed by the above (B.2) characterization of the GEM space X. We now show that the natural inclusion $\mathbf{CW}_A X \dashrightarrow SP^\infty \mathbf{CW}_A X$ admits a homotopy left inverse $SP^\infty \mathbf{CW}_A X \longrightarrow \mathbf{CW}_A X$. This left inverse will be given here as a composition

$$SP^\infty \mathbf{CW}_A X \xrightarrow{\tau^\infty} \mathbf{CW}_A SP^\infty X \xrightarrow{\mathbf{CW}_A r} \mathbf{CW}_A X,$$

where the map on the right is $\mathbf{CW}_A$ applied to the given retraction r, and the commutation map τ^∞ is given as follows:

B.5 CLAIM: *For any $0 \leq k \leq \infty$ there is a natural transformation in hoS_*,*

$$\tau^k \colon SP^k \mathbf{CW}_A \longrightarrow \mathbf{CW}_A SP^k.$$

These transformations are compatible in k.

Proof: For any given space X the map $\tau^k : SP^k \mathbf{CW}_A X \longrightarrow \mathbf{CW}_A SP^k X$ is given as follows: If k is finite we notice that, since $SP^k Y$ is Y-cellular for all Y and since $\mathbf{CW}_A X$ is A-cellular, we get that $SP^k \mathbf{CW}_A X$ is A-cellular. Therefore by universality of $\mathbf{CW}_A$ a map τ^k is given, uniquely up to homotopy, by the factorization of the map $SP^k \mathbf{CW}_A X \longrightarrow SP^k X$, which we take as $SP^k(\mathbf{CW}_A X \longrightarrow X)$. Since all maps we took are natural or uniquely determined by such up to homotopy, and are compatible with the inclusions $SP^k X \hookrightarrow SP^{k+1} X$, the claim follows.

Now the claim of (B.3) is a special ($k = \infty$) case of the following

CLAIM: *For any $0 \leq k \leq \infty$, if X is a homotopy retract of $SP^k X$ then so is $\mathbf{CW}_A X$.*

Proof: The map τ^k just defined fits into a homotopy commutative diagram as follows:

(B.5.1)

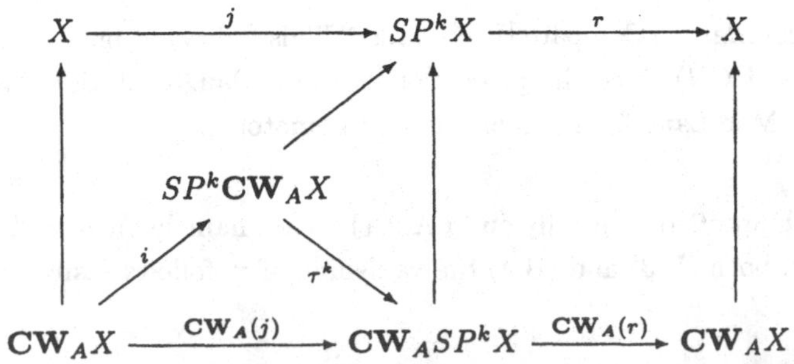

Notice, e.g., that the two compositions $\mathbf{CW}_A(r) \circ \tau^k \circ i$ and $\mathbf{CW}_A(r) \circ \mathbf{CW}_A(j) = \mathbf{CW}_A(\mathrm{id}) = \mathrm{id}$ are homotopic, since they are homotopic after composition into X and factorization of (2.E.8) is unique up to homotopy. For a similar reason $\tau^k \circ i \sim \mathbf{CW}_A(j)$. Thus i has a left inverse as needed, since $\mathbf{CW}_A(r) \circ \mathbf{CW}_A(j) \sim 1$.

To continue with the proof, since we assume the vanishing of the homotopy groups above dimension N,the vanishing of higher homotopy groups of $\mathbf{CW}_A X$ is now a simple consequence of universality or idempotency: Since the augmentation map to X is universal, if it is null on any retract W of $\mathbf{CW}_A X$, then $W \simeq *$. Compare (1.C.4) above. Since we now know that $\mathbf{CW}_A X$ is a GEM, every homotopy group of this space is carried by an appropriate Eilenberg–Mac Lane space that is a retract of $\mathbf{CW}_A X$ itself. But all maps from higher Eilenberg–Mac Lane spaces to our X are null, since by assumption this GEM is concentrated in lower dimensions. Hence these higher groups must vanish as needed. Dually, it follows that the lower homotopy groups of $\mathbf{L}_f X$ vanish, as needed.

Now the proof for the augmented functor $\overline{\mathbf{P}}_A$ is similar using universality and the consequence of (A.2.1), that if $\overline{\mathbf{P}}_A$ kills a space X it also kills any symmetric product of X. Since the above proof constructs a natural retraction, it presents by the same token the augmentation map as a retract of the symmetric product on itself.

Proof of B.4: The proof here is similar to the proof of (B.3) above. We will only give the retraction $R : SP^k \mathbf{L}_f X \longrightarrow \mathbf{L}_f X$ for any $0 \le k \le \infty$, given a retraction

$$X \longrightarrow SP^k X \xrightarrow{r} X.$$

Again R is given as a composition

$$SP^k \mathbf{L}_f X \xrightarrow{\tau^k} \mathbf{L}_f SP^k X \xrightarrow{l(r)} \mathbf{L}_f X.$$

Here τ^k is defined as a composition

$$SP^k \mathbf{L}_f X \xrightarrow{i} \mathbf{L}_f SP^k \mathbf{L}_f X \simeq \mathbf{L}_f SP^k X.$$

We now explain the homotopy equivalence. This is not immediate since as it stands SP^k is a direct limit construction, not a homotopy direct limit.

B.6 PROPOSITION: *For any connected space X there is an equivalence which is natural up to homotopy:*

$$\mathbf{L}_f SP^k \mathbf{L}_f X \simeq \mathbf{L}_f SP^k X.$$

Proof: This would be a direct consequence of (1.D.3) above once we present SP^kY as a homotopy colimit of a canonical diagram associated with Y and the integer k. But this presentation is given in the proof of (A.2) and (A.2.1) above: Thus $SP^kY = Y^k/S_k$ is given as a hocolim_* of a diagram consisting of finite products Y^ℓ for $0 \le \ell \le k$. Since $\mathbf{L}_f$ commutes with products up to homotopy (1.A.8) (e.4), we can write SP^kX and $SP^k(\mathbf{L}_f X)$ as homotopy colimits of similarly shaped diagrams of partial products X^ℓ and $(\mathbf{L}_f X)^\ell$, as needed in (1.D.3). Compare (A.2.3) above.

Proof of B.4.1: For the case of the nullification functor $\mathbf{P}_A$, this result is just a formal consequence of the long exact sequence of the homotopy groups of fibration sequence $\overline{\mathbf{P}}X \to X \to \mathbf{P}X$ using the vanishing of the higher homotopy groups of the fibre and the vanishing of the lower homotopy groups of the base as given in (B.3) and (B.4). In the general case it follows directly from the fact that not only the localization and cellularization $\mathbf{CW}_A$ of a GEM are GEMs, i.e. homotopy retracts of their own infinite symmetric power SP^∞, but the augmentation map is a GEM map as given by the following diagram:

(B.6.1)
$$\begin{array}{ccccc} X & \longrightarrow & SP^\infty X & \longrightarrow & X \\ \downarrow & & \downarrow & & \downarrow \\ \mathbf{L}_f X & \longrightarrow & SP^\infty \mathbf{L}_f X & \longrightarrow & \mathbf{L}_f X \end{array}$$

C. Relation between colimits and pointed homotopy colimits

In this section we outline a $\Sigma^2 A$-lower estimate for the cofibre of the natural comparison map $\int_I \underset{\sim}{X} \longrightarrow \text{colim}_I \underset{\sim}{X}$, in terms of the 'fixed points' spaces in X, i.e. $\text{map}_I(e, \underset{\sim}{X})$ for the orbits e that occur in $\underset{\sim}{X}$. This is carried along the lines of (A.5) above.

Recall [DF-1] that an orbit e over a small category I is an I-diagram e with $\text{colim}_I(e) \simeq *$, i.e. whose strict colimit is a point. For $I = G$ a group, this is the usual definition. In [DF-1] we associated in the usual way, with any small category O of I-orbits and any I-diagram $\underset{\sim}{X}$, an O^{op}-diagram $\underset{\sim}{X}^O$ with $\underset{\sim}{X}^O(e) = \text{map}_I(e, \underset{\sim}{X})$.

If O is the set for all orbits $\underset{\sim}{X}$, namely all the pullbacks $pt \xrightarrow{u} \operatorname{colim}_I \tilde{X} \longleftarrow\!\!> \tilde{X}$ for all u, then the associated O-diagram X^O is *free* over O^{op}.

C.1 EXAMPLE Here is a generic example: Let I be the 'pushout' category $\cdot \longleftarrow \cdot \longrightarrow \cdot$ and consider the following I-space: $\underset{\sim}{X} = \{\{1\} \longleftarrow \{0,1\} \longrightarrow \{1\}\}$. Then the comparison map from the homotopy colimit to the strict colimit is

$$S^1 \longrightarrow (pt)$$

and the cofibre is S^2. Since $\underset{\sim}{X}$ itself is an orbit $\operatorname{map}_I(\underset{\sim}{X}, \underset{\sim}{X})$ is a discrete space with few points. Hence in this case the statement of Theorem (C.8) below is $S^2 \vee S^2 \vee S^2 \ll S^2$, which is clearly correct.

C.2 POINTED, UNPOINTED DIAGRAMS. In general, the diagram $\underset{\sim}{pt}: I \longrightarrow S_*$ with $\underset{\sim}{pt}(\alpha) = *$ a single point for all $\alpha \in obj(I)$ is not a free diagram — it has no free generator. Think of a group acting trivially on one-point space; see Appendix HL below. In the pointed category, however, every (pointed) diagram has $\underset{\sim}{pt}$ as an orbit, and a *free* diagram in S_*^I is one that is free *away* from that orbit. In other words, all the orbits except the orbit of base points are free. Now if $\underset{\sim}{X}$ is any pointed I-diagram, then of course $\{\operatorname{map}_I(e, \underset{\sim}{X})\}_{e \in O}$ will also be pointed since, for each $e \in O$, there is a single map $e \longrightarrow \underset{\sim}{pt}$ and so the map $e \longrightarrow \underset{\sim}{pt} \longrightarrow \underset{\sim}{X}$ gives us a base point in $\underset{\sim}{X}(e)$ which turns $\underset{\sim}{X}^O$ into a *free pointed* diagram over O.

In the discussion below we restrict our attention to diagrams $\underset{\sim}{X}$ in S_*^I, the pointed category. Thus 'free' means free in the pointed sense. Recall the definition of pointed homotopy limit: We start with an arbitrary pointed I-space (I-diagram) and resolve it by a pointed-free I-space $E\underset{\sim}{X} \longrightarrow \underset{\sim}{X}$, with $E\underset{\sim}{X}_\alpha \longrightarrow \underset{\sim}{X}_\alpha$ a pointed homotopy equivalence for all $\alpha \in I$. Then the strict colimit of $E\underset{\sim}{X}$ is the homotopy colimit of $\underset{\sim}{X}$:

$$\operatorname{colim}_I E\underset{\sim}{X} \equiv \int \underset{\sim}{X}.$$

In [B-K], we get $\int_I \underset{\sim}{X}$ as a functor by taking a *canonical* pointed free resolution of $\underset{\sim}{X}$.

C.3 PRESENTATION OF STRICT COLIMIT AS A HOMOTOPY COLIMIT: Let e_α be the free orbit generated at α. Then $\operatorname{map}_I(e_\alpha, \underset{\sim}{X}) = \underset{\sim}{X}_\alpha$. Now it is immediate from the definition of orbit e that

C.3.1 PROPOSITION: *If $O = O(\underset{\sim}{X})$, then*

$$\operatorname*{colim}_{O^{op}} \underset{\sim}{X}^{O} \simeq \operatorname*{colim}_{I} \underset{\sim}{X}$$

is a natural equivalence.

Proof: The free orbits e_α are in O.

The main advantage of the presentation of colim_I as $\operatorname{colim}_{O^{op}}$ is that, for a free O^{op}-diagram $\underset{\sim}{X}^{O}$, the latter is also the homotopy limit:

C.4 PROPOSITION: *If $O(\underset{\sim}{X}) = O$, then the maps*

$$\operatorname*{colim}_{I} \underset{\sim}{X} \simeq \operatorname*{colim}_{O^{op}} \underset{\sim}{X}^{O} \simeq \operatorname*{hocolim}_{*\, O^{op}} \underset{\sim}{X}^{O}$$

are natural homotopy equivalences.

Proof: Under the assumption $\underset{\sim}{X}^{O}$ is a pointed free O^{op}-diagram, so its $\operatorname{colim}_O \underset{\sim}{X}$ and $\operatorname*{hocolim}_{*\,O} \underset{\sim}{X} \simeq \int_O \underset{\sim}{X}$ are equivalent by definition.

Cofibre of the comparison map. The estimate for the cofibre of the comparison map follows from a formula for it as a pointed homotopy co-equalizer, and thus pointed homotopy colimit.

Let I be a small index category and let O be a set of I-orbit $O = \{e\}$ with $e\colon I \longrightarrow S_*$, an orbit, and $\operatorname{colim}_I e = (pt)$.

An I-space $X\colon I \longrightarrow S_*$ is called of orbit-type O if it is built only from orbits in O. We assume that for all $\alpha \in I$, the free orbit e_α generated at $\alpha \in I$, is in O. We denote by X^O the associated O^{op}-space. In addition, we consider $\overline{W}O$ an O-diagram defined as follows.

C.5 DEFINITION: *For all $e \in O$, $\overline{W}O(e) = \operatorname*{hocolim}_I e$ (free hocolim). Since hocolim is a functor we get a functor $\overline{W}O\colon O \longrightarrow S$.*

Notice that we used *free* hocolim and thus $\overline{W}O$ is an *unpointed* diagram.

For example, if $I = G$ a group and O is a category of orbits $O = \{G/H\}_{H\subset G}$, then

$$\overline{W}O(G/H) = \operatorname*{hocolim}_G G/H \simeq EG \times_G G/H = BH.$$

With this definition, we can formulate the main technical result. This result expresses the cofibre of the comparison map between a strict direct limit and the corresponding homotopy direct limit in terms of a certain homotopy coend where we use the double integral notation from (A.5.1) and Appendix HC.

C.6 THEOREM: *For any pointed I-diagram $\underset{\sim}{X}$ of type $O = O(\underset{\sim}{X})$, one has a natural homotopy equivalence:*

$$\mathrm{cof}(\mathrm{hocolim}_* \underset{\sim}{X} \longrightarrow \underset{*}{\mathrm{colim}}\, \underset{\sim}{X}) \simeq \iint_O \overline{W}O * \underset{\sim}{X}^O$$

where $*$ denotes the reduced joint between unpointed and pointed spaces.

C.7 REMARK: If A is unpointed and B is pointed, one can consider $A * B$ as the pointed reduced join which is homotopy equivalent to the usual join and is obtained by collapsing the subjoin $A * \{b\}$, with $b \in B$ being the base point ($A * \{b\} \simeq \mathrm{Cone}(A)$). We denoted by $\iint_O$ the homotopy coend over O, which is a special case of homotopy colimit.

C.8 THEOREM: *Under the conditions of (C.6) above, if all the spaces in $\underset{\sim}{X}^O$ are A-cellular then the cofibre* $\mathrm{cof}(\mathrm{hocolim}_* \underset{\sim}{X} \longrightarrow \mathrm{colim}_* \underset{\sim}{X})$ *is $\Sigma^2 A$-cellular.*

Proof: Since the coend in the theorem is also a pointed homotopy colimit, in order to show that it is constructed out of $\Sigma^2 A$ it is sufficient to show that for each $(e, e') \in O^{op} \times O$, the space $Be * X^{e'} \gg \Sigma^2 A$. But since $\mathrm{colim}_I e$ is a single point Be, the nerve of e is connected so $S^1 \ll Be$ and, by assumption, $A \ll X^{e'}$ for all $e' \in 0$. Therefore $\Sigma^2 A = S^1 * A \ll Be * X^{e'}$, as needed.

Proof of Theorem C.6: The proof is analogous to the above computation (A.6) and we omit it here.

D. Application: Cellular version of Bousfield's key lemma

One formulation of the lemma is as follows:

D.1 PROPOSITION: *Let X be a connected space. For any integer $k > 0$ the suspension of the Dold–Thom map $\Sigma X \longrightarrow \Sigma SP^k X$ induces a homotopy equivalence*

$$\mathbf{P}_{\Sigma^2 X}\Sigma X \xrightarrow{\simeq} \mathbf{P}_{\Sigma^2 X}\Sigma SP^k X$$

D.2 REMARK: This innocent-looking lemma which is due to Bousfield [B-4] is really a key technical result on which much of the present development rests.

The statement is equivalent in less technical terms to the following statement originally given by Bousfield: If $\mathrm{map}_*(\Sigma^2 X, Y) \simeq *$ for Y a 1-connected space, then the Dold–Thom map induces an equivalence on function complexes $\mathrm{map}_*(X, \Omega Y) \simeq$

$\mathrm{map}_*(SP^kX, \Omega Y)$ for all k, or equivalently $\mathrm{map}(\Sigma X, Y) \cong \mathrm{map}(\Sigma SP^k X, Y)$. To see the equivalence note that the condition on Y is that it is $\Sigma^2 X$-null, so $\mathrm{map}_*(T, Y) \simeq \mathrm{map}(\mathbf{P}_{\Sigma^2 X} T, Y)$ for all $T \in \mathcal{S}_*$, hence if we assume (D.1) we get the equivalence of the function complexes into Y. On the other band the statement given by Bousfield guarantees that the suspension of the Dold–Thom map on X is $\Sigma^2 X$-null equivalence (1.C.5) which is precisely what (D.1) claims again by (1.C.5).

Proof: This is based on [Dw-2]. The key element behind the present proof is Theorem A.5 above. One considers SP^kX as the quotient space $\mathrm{colim}_{S_k} X^k$ and we notice that X itself is equal to the quotient space: $\mathrm{colim}_{S_k} \bigvee_k X$. The map $X \longrightarrow SP^kX$ is induced by $\bigvee_k X \to \prod_k X$ by taking colim_{S_k}, where S_k is the symmetric group on k-letters. So the proposition is translated via the S_k equivariant map:

$$\Sigma\sigma_k\colon \Sigma \operatorname*{colim}_{S_k} \bigvee_k X \longrightarrow \Sigma \operatorname*{colim}_{S_k} \Pi_k X$$

to the statement $\mathbf{P}_{\Sigma^2 X}(\Sigma\sigma_k)$ is a homotopy equivalence.

The idea of the proof is to notice that had these colimits been (pointed) homotopy colimits, then the result would have been immediate. Since the cofibre of the map of suspensions *before* taking these hocolimits is clearly $\Sigma^2 X$-cellular—see below—it would have remained so after taking pointed hocolims. The body of the proof uses (A.5) to circumvent the difficulty by comparing the colimits with the homotopy colimits of the same spaces over S_k which are pointed with respect to the action.

Let us begin with the

Observation: $\Sigma\eta_k\colon \Sigma \bigvee_k X \longrightarrow \Sigma\prod_k X$ induces a homotopy equivalence $\mathbf{P}_{\Sigma^2 X}\Sigma\eta_k$ upon taking the $\Sigma^2 X$-nullification.

Let us prove this for the case $k = 2$. We get $\Sigma(X \times X) = \Sigma X \vee \Sigma X \vee X * X$. Since X is connected, we have $S^1 \ll X$, so $S^1 * X \ll X * X$ or $\Sigma^2 X \ll X * X$.

Therefore we can compute using (1.D.2), (1.D.5)(4) above:

$$\mathbf{P}_{\Sigma^2 X}\Sigma(X \times X) \simeq \mathbf{P}_{\Sigma^2 X}(\Sigma X \vee \Sigma X \vee X * X) \simeq$$
$$\simeq \mathbf{P}_{\Sigma^2 X}(\mathbf{P}_{\Sigma^2 X}(\Sigma X \vee \Sigma X) \vee \mathbf{P}_{\Sigma^2 X} X * X) \simeq \mathbf{P}_{\Sigma^2 X}(\Sigma X \vee \Sigma X).$$

For $k \geq 2$ the argument is identical involving the decomposition of suspension of the higher powers of X as a wedge sum.

We now compare in the following diagram wherein all hocolims are over S_k. We claim that upon application of $\mathbf{P}_{\Sigma^2 X}$ to diagram D.2 it becomes a diagram of weak equivalences so at the bottom row we get the claim of (D.1) as needed. The vertical maps $\Sigma c_1, \Sigma c_2$ are the suspensions of the natural comparison maps from the pointed Borel constructions to the orbit spaces as in (A.5) above. The pointed homotopy colimits appearing below and denoted here by $\int$ are all pointed Borel constructions (with respect to the symmetric group on k letters) or suspensions thereof:

$$
\text{(D.3)}\qquad
\begin{array}{ccc}
\int \Sigma \bigvee_k X & \xrightarrow{\int \Sigma\eta_k} & \int \Sigma \Pi_k X \\
\simeq \downarrow g_1 & & \simeq \downarrow g_2 \\
\Sigma \int \bigvee_k X & \xrightarrow{\Sigma \int \eta_k} & \Sigma \int \Pi_k X \\
\downarrow \Sigma c_1 & & \downarrow \Sigma c_2 \\
\Sigma \operatorname{colim}_{S_k} \bigvee_k X & \xrightarrow{\Sigma\sigma_k} & \Sigma \operatorname{colim}_{S_k} \Pi_k X \\
= \downarrow \mathrm{id} & & = \downarrow \mathrm{id} \\
\Sigma X & \xrightarrow{\Sigma c} & \Sigma SP^k X
\end{array}
$$

By the observation above the map $\mathbf{P}_{\Sigma^2 X}\Sigma\eta_k$ is a homotopy equivalence. We will now show using standard properties of pointed homotopy colimit ($\int$) that this implies:

CLAIM: The top horizontal map in the ladder (D.3) namely, $\int \Sigma\eta_k$, induces an equivalence upon applying $\mathbf{P}_{\Sigma^2 X}$ to it.

Proof of claim: Since $\Sigma\eta_k$ is an equivariant map we consider it as a map of diagrams and apply to it Proposition 1.D.2 to show that $\int \Sigma\eta_k$ also induces a homotopy equivalence on $\mathbf{P}_{\Sigma^2 X}$:

$$\mathbf{P}_{\Sigma^2 X} \int \Sigma\eta_k \simeq \mathbf{P}_{\Sigma^2 X} \int \mathbf{P}_{\Sigma^2 X}\Sigma\eta_k$$

where we have applied the comparison (1.D.2) to the map $\Sigma\eta_k$.

In other words, we have the following two special cases of (1.D.2):

$$\mathbf{P}_{\Sigma^2 X}\Big(\int \Sigma \bigvee_k X\Big) \simeq \mathbf{P}_{\Sigma^2 X}\Big(\int \mathbf{P}_{\Sigma^2 X}\Sigma \bigvee_k X\Big)$$

and

$$\mathbf{P}_{\Sigma^2 X}\Big(\int \Sigma\Pi_k X\Big) \simeq \mathbf{P}_{\Sigma^2 X}\Big(\int \mathbf{P}_{\Sigma^2 X}\Sigma\Pi_k X\Big).$$

Now the observation above shows the the the two spaces on the right of these equivalences are equivalent since it says that before applying to both $\mathbf{P}_{\Sigma^2 X} \int$ they were already equivalent via an equivariant map $\mathbf{P}_{\Sigma^2 X}\Sigma\eta_k$–since nullification is functorial it turns an equivariant map into such. But taking homotopy colimit always turns an equivariant map that is an equivalence into an equivalence. Therefore $\int \mathbf{P}_{\Sigma^2 X}\Sigma\eta_k$ and thus also $\mathbf{P}_{\Sigma^2 X}\int \mathbf{P}_{\Sigma^2 X}\Sigma\eta_k$ are equivalences. We conclude from this that $\int \Sigma\eta$ induces an equivalence on $\mathbf{P}_{\Sigma^2 X}$ as claimed.

By commutation of pointed homotopy with suspension $\Sigma\int \eta_k \sim \int \Sigma\eta$ induces homotopy equivalence upon applying $\mathbf{P}_{\Sigma^2 X}$. This takes care of the top two arrows in figure (D.3) above. Notice that g_1, g_2 are homotopy equivalences, again since suspension commutes with *pointed* hocolim$_*$.

The last point of the proof is now to show that the two vertical comparison maps comparing the pointed Borel constructions with the strict direct limits again induce equivalence upon applying $\mathbf{P}_{\Sigma^2 X}$. To show that $\mathbf{P}_{\Sigma^2 X}$ applied to Σc_i is a weak equivalence we use (1.D.4) and (1.D.5) (5) above: As we shall immediately see it is enough to show that $\mathbf{P}_{\Sigma^2 X}\mathrm{Cof}(c_i) \simeq *$ (where Cof(-) denotes the cofibre–i.e. mapping cone—of a map) and then use the usual Puppe sequence. For these we claim that both cofibres are $\Sigma^2 X$-cellular. In fact we claim that c_1, c_2 satisfy the conditions of (A.5) and thus by (A.5) $\mathbf{P}_{\Sigma^2 X}\mathrm{Cof}(c_i) \simeq *$ which implies that $\mathbf{P}_{\Sigma^2 X}(\Sigma c_1)$ and $\mathbf{P}_{\Sigma^2 X}(\Sigma c_2)$ are homotopy equivalences. The map c_2 is exactly the map appearing in (A.5.1) so the inequality $\Sigma^2 X \ll \mathrm{Cof}(c_i)$ holds. The same reasoning as in (A.5.1) holds for the map c_1 since again the fixed points spaces of the action of the permutation group are X-cellular: Notice that the orbit category of the symmetric group is $\{S_k/H\}_{H\subseteq S_n}$ and the spaces $(\Pi_k X)^{S_k/H}$ are partial products $X_{i_1} \times X_{i_2} \times \cdots \times X_{i_l}$ while the spaces $(\bigvee_k X)^{S_k/H}$ are always of the form: $X_{i_1} \vee X_{i_2} \vee \cdots \vee X_{i_m}$ because of the nature of the action. Therefore, all these fixed-point spaces with respect to any subgroup are in fact X-cellular and connected. Therefore, in the cofibration Puppe sequence

$$\int_{S_k} \Pi_k X \xrightarrow{c_1} SP^k X \longrightarrow C_1 \longrightarrow \Sigma \int_{S_k} \Pi X \xrightarrow{\Sigma c_2} \Sigma SP^k X$$

we have $\Sigma^2 X \ll C_1$, so $\mathbf{P}_{\Sigma^2 X} C_1 \simeq *$ and thus the induced map $\mathbf{P}_{\Sigma^2 X}(\Sigma c_2)$ is a homotopy equivalence. A similar argument works for C_2, where we replace ΠX by the wedge $\bigvee X$ and $SP^k X$ by X; we get that the map $\mathbf{P}_{\Sigma^2 X}(\Sigma c_1)$ is a homotopy equivalence. To conclude, we saw, using standard $\mathbf{P}_A$-calculus (1.A.8) and Theorem A.5, that all the maps in the above comparison ladder induce equivalence on $\mathbf{P}_{\Sigma^2 X}$, therefore so do the maps $\Sigma\sigma_k$ and Σc as claimed.

∎

The following proposition formulates a weak 'geometric' aspect of the key lemma above. It show that to construct the suspension of the symmetric product of X out of the suspension of X one needs precisely one copy of ΣX and then one can proceed by attaching copies of higher suspensions $\Sigma^{2+n}X$ of X:

D.4 PROPOSITION: *Let X be a connected space. For any integer $k \geq 0$ the suspension of the Dold–Thom map $\Sigma X \longrightarrow \Sigma SP^k X$ gives cellular inequality:*

$$\Sigma^2 X \ll \Sigma(SP^k X/X).$$

Proof: This follows easily from diagram D.3 above by taking cofibres of the maps involved.

5. GENERAL THEORY OF FIBRATIONS, GEM ERROR TERMS

Introduction

In the present chapter we consider the behaviour of fibration sequences under $\mathbf{L}_f, \mathbf{P}_A$ and $\mathbf{CW}_A$ as well as some cofibration sequences. In particular, in continuation of the work in Chapter 3, more precise information about the way these functors commute with the loop space functor and other mapping spaces lies at the center of our interest. We also consider briefly commutation with suspensions. We would also like to relate, as closely as possible, $\mathbf{L}_f\text{map}(V, X)$ and $\text{map}(V, \mathbf{L}_f X)$. The main steps in this direction were taken in [B-4] and were further developed in [DF-S]. It turns out that if A or f are suspensions or double suspensions, then these functors come very close to preserving general fibration sequences over connected spaces. In practice, we must first consider the functor $\mathbf{P}_A$. One can show that, short of a 'small abelian error term', the nullification functor $X \to \mathbf{P}_A X$ with respect to any suspension $A = \Sigma A'$ preserves homotopy fibration sequences. Explicitly, for any map g, the homotopy fibre of the natural map from the nullification of the homotopy fibre of g to the fibre of its nullification is always a GEM (see 4.B.1) for such an A. More generally, the f-localization functor $X \to \mathbf{L}_f X$ with respect to double suspensions $f = \Sigma^2 f'$ preserves homotopy fibration sequences up to a 'generalized Postnikov stage'—polyGEM. This implies as a special case that Bousfield's homological localization functor $\mathbf{L}_E$ with respect to any general homology theory E_* preserves fibrations of the form $\Omega^2 F \to \Omega^2 E \to \Omega^2 B$ up to polyGEM. Moreover, for p-torsion theories such as mod-p K-theory and higher Morava K-theories, the error term is rather small and lives in three dimensions typical of the given theory. Let us quote here two results from Section B below that are typical. These are dual results, one dealing with fibrations the other with cofibrations.

MAIN THEOREM B.1: *Let $f : A \to D$ be a map of connected spaces. If $F \to E \to B$ is any fibration with $\mathbf{L}_{\Sigma f} E \simeq \mathbf{L}_{\Sigma f} B \simeq *$ then $\mathbf{L}_{\Sigma f} F$ has a natural structure of a GEM.*

Dually for cofibration one has:

COROLLARY B.4.1: *Let $\Sigma A \to \Sigma X \to \Sigma(X/A)$ be a suspension of a cofibration sequence. If $\mathbf{L}_f \Sigma X \simeq \mathbf{L}_f \Sigma(X/A) \simeq *$ then the localization $\mathbf{L}_f \Sigma A$ has a natural structure of a GEM.*

This in turn has an interesting implication on the partial order $(<)$ defined in (1.A.5):

COROLLARY B.6: *For any spaces X, Y and integer $k \geq 1$ one has: $\Sigma X < \Sigma Y$ iff $\Sigma^k X < \Sigma^k Y$.*

Proofs: The proof of (B.1) occupies most of Section B below. The two corollaries (B.4.1) and (B.6) are treated at the end of Section B.

Remark: We see that if $\mathbf{L}_{\Sigma f}$ kills both base and total space of a fibration sequence, then it turns the fibre in a natural way into an infinite loop space that is equivalent as such to a product of (abelian) Eilenberg–Mac Lane spaces equipped with their standard infinite loop space structure. Thus $\mathbf{L}_{\Sigma f}$ 'abelianizes' the fibre, killing in the process all its k-invariants (and presumably much of its homotopy groups, too). In fact, in many important cases under the above assumption $\mathbf{L}_f F$ is just a $K(G, n)$ for some abelian group G and integer n. Notice that by (1.H.1) above and in contrast to (B.1), if $\mathbf{L}_f$ kills any *other* pair of spaces in $F \to E \to B$ then it kills the third too, and here f need not be a suspension. In a suspension of any cofibration sequence similar behaviour occurs.

Example: A typical example of the situation depicted above occurs when one applies the Postnikov section functor $P_n = \mathbf{P}_{S^{n+1}}$ to a fibration of the form $\Omega X \to * \to X$. If X is n-connected, then $P_n X \simeq *$ but $P_n \Omega X$ is equivalent to an Eilenberg–Mac Lane space $K(\Pi, n)$ with Π the first non-trivial homotopy group of X. Similarly, if a Postnikov section of a double suspension of any space is contractible then the same section of the single suspension must be an Eilenberg-Mac Lane space. It is surprising that, despite its simplicity, this is the generic situation. It motivated Bousfield in the first place to look and find his early, most important, version of (B.1.1) below.

The control gained on the behaviour of fibration sequences under localization will allow us to prove Theorem 6.A.1 (in the next chapter) which gives information about homological localizations of double-loop fibre sequences.

A. GEM and polyGEM error terms – main results

In this section we state most of the main results about preservation of fibrations by homotopy localization functors. Since some of the proofs are a bit lengthy we will give these proofs together with several interesting additional propositions in sections B, C and D below.

A.1 DEFINITION: *We define the class of (twisted) polyGEMs (respectively, oriented polyGEMs) as follows: It is the smallest full sub-category that includes all the underlying spaces of GEM spaces as members and is closed under taking the total space (respectively, homotopy fibre) of any fibration sequence for which the other two members are polyGEMs.*

Thus an oriented polyGEM is built from GEMs by a finite number of oriented, principal fibrations while in a general twisted polyGEM arbitrary fibrations are allowed. In particular, oriented polyGEMs are all nilpotent spaces whereas in general (twisted) polyGEMs the action of the fundamental group on the higher homotopy groups is not restricted, while the fundamental group itself can be any solvable group.

The class of oriented polyGEMs is filtered naturally as follows:

A 1-polyGEM is the underlying space of

a GEM. If $W_1 \to W_2$ is a map of k-polyGEMs, then the homotopy fibre is a $(k+1)$-polyGEM.

A.1.1 EXAMPLE: Any finite Postnikov section of a space with abelian fundamental group is a polyGEM and if it is nilpotent as a space [B-K] then it is an oriented polyGEM. In the present context compare (F.7) below.

A.2 DEFINITION: *We say that the coaugmented functor $X \to \mathbf{L}X$ preserves the fibre sequence $F \to E \to B$ over the pointed space B up to a (poly)GEM error term if one can construct a map c into the homotopy fibre of $\mathbf{L}E \to \mathbf{L}B$,*

$$c : \mathbf{L}F \to \text{fibre}(\mathbf{L}E \to \mathbf{L}B),$$

whose homotopy fibre $J = \text{Fib}(c)$ is a (poly)GEM (see A.1).

A.3 EXAMPLE: For $\mathbf{L} = \mathbf{L}_f$, the homotopy f-localization functor, one can associate with any fibration $F \to E \to B$ over a pointed space B a map $\mathbf{L}F \to$ fibre $(\mathbf{L}E \to \mathbf{L}B)$ that is natural up to homotopy:

This *comparison map* is given by universality of $\mathbf{L} = \mathbf{L}_f$ and the fact (1.A.8)(e.3) above, that the space fibre $(\mathbf{L}E \to \mathbf{L}B)$ is f-local. For the standard loop space fibration $\Omega X \to * \to X$ the above map is the 'commutation map' $\mathbf{L}\Omega X \to \Omega\mathbf{L}X$.

Let us now formulate the three main consequences of Theorem B.1 quoted above when coupled with results of Chapter 3: The following is an interesting special case of (A.5) given below when applied to the path fibration over X:

A.4 THEOREM: *Let $A = \Sigma A'$ be any suspension space and let X be any simply connected space. Then the fibre J of the natural map σ,*

$$\sigma : \mathbf{P}_A\Omega X \to \Omega\mathbf{P}_A X,$$

is a GEM. In other words, $\mathbf{P}_A$ commutes with Ω up to a GEM. Moreover, J is A-null and $\mathbf{P}_{A'} J \simeq \{*\}$.

∎

Remark: For a given space A' of finite type it is a simple matter to determine the local $\Sigma A'$-null GEM's which localize to $\{*\}$ under $\mathbf{P}_{A'}$.

A.5 THEOREM: *Let $A = \Sigma A'$ be any suspension. Then $\mathbf{P}_A$ preserves any fibration sequence up to a GEM error term J. In fact, $\mathbf{P}_{A'}$ preserves up to a GEM any loop fibration $\Omega F \to \Omega E \to \Omega B$ for any space A'. This GEM error term J is A-null and $\mathbf{P}_{A'}$ kills it: $\mathbf{P}_{A'} J \simeq *$.*

Proof: The proof is given in Section C.3 below.

Remark: A further interesting property of J is explained in (A.10) below.

And finally,

A.6 THEOREM: *For any map $g : A \to D$ the functor $\mathbf{L}_g$ preserves the double loops of any fibration sequence: $\Omega^2 F \to \Omega^2 E \to \Omega^2 B$, up to a polyGEM error term $\Omega^2 J$ which is g-local and satisfies $\mathbf{L}_g J \simeq *$.*

Proof: The proof is given in Section D below, see (D.6)(1),(2),(3).

If one applies (A.6) to the standard loop space fibration one gets:

A.7 COROLLARY: *When localizing a connected pointed space X with respect to any double suspension map, the homotopy difference between the localization of the loop of X and the loop of the localization of X is always a polyGEM.* ∎

A.8 Homotopically discrete mapping spaces

In Theorem A.5 we saw that the error term is intuitively 'small' since it is $\Sigma A'$-null but it is killed by $\mathbf{P}_{A'}$. such spaces arise naturally here and they have a very interesting and useful property noticed in certain cases in [B-4]:

A.9 PROPOSITION: *if Y is a ΣW-null space that is killed by $\mathbf{P}_W$ i.e. with $\mathbf{P}_W Y \simeq *$ then the space of pointed self maps $Y^Y = \mathrm{map}_*(Y, Y)$ has a contractible null-component.*

Proof: The proof is easy: All we need to show is that the loop space of the mapping space is contractible. But we have

$$\Omega \,\mathrm{map}_*(Y,Y) \simeq \Omega \,\mathrm{map}_*(Y, \mathbf{P}_{\Sigma W} Y) \simeq \mathrm{map}_*(Y, \Omega \mathbf{P}_{\Sigma W} Y) \simeq \mathrm{map}_*(Y, \mathbf{P}_W \Omega Y) \simeq *$$

The first equivalence follows from Y being ΣW-null,the third is standard commutation from Chapter 3, the last one follows from the assumption that $\mathbf{P}_W$ kills Y i.e. $Y \longrightarrow *$ is a W-local equivalence.

In the circumstances of (A.5) we see that the error term is a GEM that satisfies the conditions of (A.9) with $W = A'$. But since it is a GEM it is a loop space and all the components of $\text{map}_*(V, J)$ for any V are equivalent to each other and hence they are all contractible. Therefore we conclude from (A.5) and (A.9):

A.10 PROPOSITION: *The error term J in (A.5) is a GEM whose space of self-maps is homotopically discrete.* ∎

B. The main theorem on GEM error terms

In the present section we are concerned with the following:

B.1 THEOREM: *If $F \to E \to B$ is any fibration sequence and $f\colon A \to D$ any map of connected spaces with $\mathbf{L}_{\Sigma f}B \simeq \mathbf{L}_{\Sigma f}E \simeq *$, then $\mathbf{L}_{\Sigma f}F$ has a natural structure of a GEM.*

Remark: Moreover, it is probable that if the fibration is principal, F being a group, then the induced group structure on $\mathbf{L}_f F$ coming from (3.A.1) is homotopically equivalent to its group structure as a GEM. See diagram (B.1.4). We will not need to use this extra structure in the present notes.

B.1.1 COROLLARY: *For any map f of connected spaces, if $\mathbf{L}_{\Sigma f}X \simeq *$ then $\mathbf{L}_{\Sigma^2 f}X$ has a natural structure of a GEM.*

Remark: It might as well be assumed that the spaces in the above E, B, X are all 1-connected since this follows directly form the assumptions.

Proof: The proof of (B.1) and (B.1.1) will occupy most of the rest of this section after the following

Remarks: 1. For the case $f\colon A \to *$, i.e. $\mathbf{L}_f = \mathbf{P}_A$, and under special assumptions on A, this key result was found and proven in [B-4]. As will be clear from the following Theorem B.1 can be considered as a relative version of (B.1.1).

2. Notice the non-example for Theorem A.4 for non-suspensions: If A is an acyclic space, then $\mathbf{P}_{\Sigma A}A \simeq A$ even if $\mathbf{P}_A A \simeq *$. As a simply-connected counter-example, when A is not a suspension one can take $A = X = BS^3$. By a theorem of Zabrodsky the null component of $\text{map}_*(BS^3, BS^3)$ is contractible. Therefore BS^3 is ΣBS^3-null. So in this case $\mathbf{P}_{\Sigma A}X \simeq X$, but $\mathbf{P}_A X \simeq *$ and BS^3 is not a GEM. We conclude that one cannot dispense with some assumption on A or X in order to allow the deduction in Corollary B.1.1 below:

$$\mathbf{P}_A X \simeq * \Rightarrow \mathbf{P}_{\Sigma A}X \simeq \text{GEM}.$$

3. It is not hard to construct examples in which $\mathbf{P}_{\Sigma^2 A}(\Sigma A)$ has non-trivial homotopy in infinitely many dimensions. Take A to be the wedge of all Moore spaces

$M^p(\mathbb{Z}/p\mathbb{Z})$. $p \geq 2$; then $\mathbf{P}_{\Sigma^2 A}\Sigma A$ is the product of all $K(\mathbb{Z}/p\mathbb{Z}, p-1)$, $p \geq 2$ when p is a prime.

4. Notice the immediate interesting implication: For any connected space X the space $\mathbb{G} = \mathbf{P}_{\Sigma^2 X}\Sigma X$ has a natural structure of a GEM. This last implication does not hold for the three-point space since we get a wedge of two circles for the space $\mathbb{G}$.

5. It was shown by Casacuberta and Peschke [C-P] that the localization with respect to the degree p map between circles does not behave as nicely as in (A.4). That map, of course, is not a suspension. Still, it can be partly understood using homology with local coefficients.

6. We were unable to determine whether the polyGEM in (A.6) is in fact a GEM.

Proof of Corollary B.1.1: Since the proof here is an easier version of the proof of (B.1), we give of a direct, independent, proof of it below before that of (B.1) itself.

However, using the loop structure on localization of loop spaces it can be deduce from the remark following (B.1) as follows: We can assume that X is a connected space by looking at each component. Recall $\mathbf{L}\Omega X \simeq \Omega \mathbf{L}_{\Sigma f} X$, so our assumption implies that $\mathbf{L}_f \Omega X \simeq *$; in particular it follows that ΩX is connected and X is simply connected (1.A.3)(e.11). The deduction of the corollary is immediate from the commutation formula (3.A.1) above once the theorem is applied to the principal fibration $\Omega X \to * \to X$. The assumption $\mathbf{L}_{\Sigma f} X \simeq *$ gives, by Theorem B.1, $\mathbf{L}_{\Sigma f}\Omega X \simeq$ GEM. Therefore by (3.A.1) there is an equivalence $\Omega \mathbf{L}_{\Sigma^2 f} X \simeq$ GEM. Since these two are equivalent as abelian infinite loop spaces one can take classifying spaces using (3.A.1) to get: $\mathbf{L}_{\Sigma^2 f} X$ is also a GEM, as needed.

We now turn to an <u>independent proof of (B.1.1).</u> The first main step in the proof is:

B.1.2 LEMMA: *If* $\mathbf{L}_f \Omega X \simeq *$, *then:*

(i) The natural axis map $\mathbf{L}_{\Sigma^2 f}(X \vee X \vee \cdots \vee X) \to \mathbf{L}_{\Sigma^2 f}(X \times X \times \cdots \times X)$ *in an equivalence.*

(ii) Condition (i) implies that $\mathbf{L}_{\Sigma^2 f} X$ *has a natural infinite loop space structure.*

Proof of B.1.2 (i): For simplicity of exposition we consider only the case involving two copies of X; the general case proceeds verbatim with the obvious changes. We start our proof by recalling [G] (see examples in Appendix HL) that the homotopy fibre of the natural inclusion $i : X \vee X \to X \times X$ is homotopy equivalent to the join $\Omega X * \Omega X \simeq \Sigma(\Omega X \wedge \Omega X)$ for any pointed connected CW-complex X. Therefore, by (1.H.1), in order to show that $\mathbf{L}_{\Sigma^2 f}(i)$ is a homotopy equivalence it is sufficient to show that $\mathbf{L}_{\Sigma^2 f}(\Omega X * \Omega X) \simeq *$.

But by (1.A.8)(e.10) we always have the implication $\mathbf{L}_f A \simeq * \Rightarrow \mathbf{L}_{\Sigma f}\Sigma A \simeq *$. Since $X * Y \simeq \Sigma(X \wedge Y)$ it is sufficient to show that $\mathbf{L}_{\Sigma f}(\Omega X \wedge \Omega X) \simeq *$. For this we use(1.A.8) (e.10) above:

Since the assumption with (3.A.1) gives $\mathbf{L}_f \Omega X \simeq *$, ΩX is connected since (e.11) the map f is between connected spaces. Thus $\mathbf{P}_{S^1}\Omega X \simeq *$. Using (e.10) we get $\mathbf{L}_{S^1 \wedge f}(\Omega X \wedge \Omega Y) \simeq *$, where S^1 is the circle, as needed.

Proof of B.1.2 (ii): We proceed to show that the natural equivalence $\mathbf{L}_{\Sigma^2 f}(X \vee X \cdots \vee X) \to \mathbf{L}_{\Sigma^2 f}(X \times X \cdots \times X)$ implies that $\mathbf{L}_{\Sigma^2 f} X$ is an ∞-loop space.

For this we use [A], [S], [B-F], see also (3.C.3) above. According to Segal an infinite loop space structure on a space X is specified completely once the space is embedded in a diagram of spaces over the small category of pointed finite sets $\Gamma^+ = \{n^+ : n \geq 0\}$ (called very special Γ-space) as the space corresponding to (1^+). A *special* Γ-space is just a diagram over the category of pointed finite sets Γ^+ such that it assigns a single point to (0^+) and its value on n^+ is homotopy equivalent to the n-th power of its value on 1^+ via the obvious map which is the product of the maps $d_i : n^+ \longrightarrow 1^+$ sending every element to 0 except i that goes to $1 \in 1^+$. It is *very special* if the induced monoid structure on the path components of $\Gamma(1^+)$ is in fact an abelian group. In our case all spaces involved in the diagram are connected, in fact 1-connected, so the condition on π_0 is automatically satisfied.

Therefore, to exhibit the infinite loop space structure on the space $\mathbf{L}_{\Sigma^2 f}(X)$ as needed, we construct a (very) special Γ-space $X_\bullet$ with $X_0 \simeq *, X_1 \simeq X$. First we construct a 'non-special' Γ space $\check{X}_\bullet$ by setting $\check{X}_n = \bigvee_n X$, the point-sum of n-copies of X. If S is a pointed finite set, we really take the half-smash $\check{X}_S = X \ltimes S$ which we write, with a slight abuse of notation, as $\bigvee_S X$. Clearly for any map of finite pointed sets $S \to T$ we have a corresponding map $\bigvee_S X \to \bigvee_T X$, so $\check{X}_\bullet$ is a functor from the category of finite sets to spaces with $\check{X}_{(\text{empty})^+} = \check{X}_0 = pt$. The only condition of a 'special' Γ-space that is not satisfied by $\check{X}_\bullet$ is that the map $\check{X}_n \to \check{X}_1 \times \cdots \times \check{X}_1$ (n-times) is not a homotopy equivalence. But now we define $X_n = \mathbf{L}_{\Sigma^2 f}\check{X}_n$; since $\mathbf{L}_{\Sigma^2 f}$ is functorial we still get a Γ-space. It is special because we have the equivalence:

$$\mathbf{L}_{\Sigma^2 f}(\bigvee_S X) \simeq \prod_S \mathbf{L}_{\Sigma^2 f}\, X.$$

This follows by the natural equivalence proven above (B.1.2 (i)), together with the natural equivalence that gives the multiplicative property of $\mathbf{L}_f$ (1.A.8)(e.4) $\mathbf{L}_f(W \times W') = \mathbf{L}_f W \times \mathbf{L}_f W'$ for any f and W, W'.

This concludes the proof of Lemma B.1.2.

We now conclude the proof of (B.1.1):

By Lemma B.1.2 our space $\mathbf{L}_{\Sigma^2 f}X$ is an infinite loop space and therefore we can write $\mathbf{L}_{\Sigma^2 f}X \simeq \Omega Y$. We saw above that ΩX is connected and X is simply connected. Therefore so is $\mathbf{L}_{\Sigma^2 f}X$; we conclude that Y is simply connected.

Consider $\mathrm{map}_*(\Sigma^2 X, Y)$. We claim that it is contractible. This is true since any map $\Sigma X \to \Omega Y = \mathbf{L}_{\Sigma^2 f}X$ factors through $\mathbf{L}_{\Sigma^2 f}\Sigma X$. But since $\mathbf{L}_{\Sigma f}X \simeq *$ by assumption, the latter is equivalent to a point by (1.A.8) (e.10). Similarly, any map $\Sigma^k X \to \Omega Y$ must be null for all $k \geq 1$. Alternatively use (1.A.8) (e.9). Therefore the condition of Bousfield's key lemma (4.D.1) above and (B.2) below are satisfied, Y being simply connected. Thus any map $X \to \Omega Y = \mathbf{L}_{\Sigma^2 f}X$ factors uniquely through $\mathbf{SP}^k X$ for all $k \geq 1$. Because of the uniqueness of the factorization we can conclude that the factorizations are compatible for various k.

Therefore we get a factorization through the infinite symmetric power of X for the localization map on X:

(B.1.3)
$$\begin{array}{ccc} & & \mathbf{SP}^\infty X \\ & \nearrow & \downarrow \\ X & \longrightarrow & \mathbf{L}_{\Sigma^2 f}X \end{array}$$

If we now apply $\mathbf{L}_{\Sigma^2 f}$ to this triangle, using $\mathbf{L}_{\Sigma^2 f}\mathbf{L}_{\Sigma^2 f} \simeq \mathbf{L}_{\Sigma^2 f}$ we get that $\mathbf{L}_{\Sigma^2 f}X$ is a homotopy retract of $\mathbf{L}_{\Sigma^2}(SP^\infty X)$. But (4.B.4) asserts that $\mathbf{L}_f$ turns any GEM into a GEM. Since a retract of a GEM is a GEM we get the desired result that $\mathbf{L}_{\Sigma^2 f}X$ – being up to homotopy a retract of $\mathbf{L}_{\Sigma^2 f}SP^\infty X$ – is a GEM.

We now address the relations between the GEM structure and the infinite loop space structure. Notice that the infinite loop structures in $SP^\infty X$ and $\mathbf{L}_{\Sigma^2 f}X$ are given by definition in the same manner by the concatenations of words in the underlying spaces, so that they are naturally compatible. Explicitly, the infinite loop space structure is given in terms of special Γ-space via (B.1.2)(ii) since by the Dold-Thom theorem the axis map induces an equivalence

$$SP^\infty(X \vee X \vee \cdots \vee X) \xrightarrow{\simeq} SP^\infty X \times SP^\infty X \times \cdots \times SP^\infty X.$$

Such an equivalence yields as above an infinite loop space structure on $SP^\infty X$. Therefore the homotopy uniqueness of the vertical factorization arrow in the triangle (B.1.2) implies that we have a homotopy commutative diagram in which the horizontal arrows give the homotopy product structures on the spaces involved in view of the equivalences above:

$$\text{(B.1.4)} \qquad \begin{array}{ccc} SP^\infty(X \vee X) & \longrightarrow & SP^\infty X \\ \downarrow & & \downarrow \\ \mathbf{L}_{\Sigma^2 f}(X \vee X) & \longrightarrow & \mathbf{L}_{\Sigma^2 f} X. \end{array}$$

Therefore the GEM structure and the infinite loop space structure on $\mathbf{L}_{\Sigma^2 f}X$ are compatible. This proves (B.1.1). ∎

We now turn to the proof of the main theorem (B.1). We use the same line of argument as in the proof of (B.1.1) so details are omitted.

Proof of B.1 Referring to the proof of (B.1.2) above we first show that $\mathbf{L}_{\Sigma f}F$ is an ∞-loop space in a natural way using [S], [A]. We define a (non-special) Γ-space as follows:

$$\overset{\vee}{F}_n = \text{fibre of } (E \vee \cdots \vee E \to B \vee \cdots \vee B) \quad (n\text{-copies of } E, B).$$

The homotopy fibre being a functor in $\mathcal{S}_*$ and $\left\{\bigvee_n E\right\}_{n\geq 0} \to \left\{\bigvee_n B\right\}_{n\geq 0}$ being a map of Γ-spaces, we conclude that $\overset{\vee}{F}_\bullet$ above is a Γ-space.

<u>B.1.5 Claim:</u> *The natural map (see diagram B.1.4 below):*

$$f_n \colon \overset{\vee}{F}_n \to F \times \cdots \times F$$

induces an equivalence on $\mathbf{L}_{\Sigma f}(f_n)$ *and therefore the space* $\mathbf{L}_{\Sigma f}\overset{\vee}{F}_1 = \mathbf{L}_{\Sigma f}F$ *is an* ∞*-loop space.*

<u>Proof of Claim:</u> Since by definition $\check{F}_1 = F$ the equivalence in the claim implies that:

$$\mathbf{L}_{\Sigma f}\check{F}_n \simeq \mathbf{L}_{\Sigma f}(F\times\cdots\times F) \simeq (\mathbf{L}_{\Sigma f}F)^n \simeq (\mathbf{L}_{\Sigma f}\check{F}_1)^n.$$

Thus $\mathbf{L}_{\Sigma f}(\check{F}_\bullet)$ is a *special* Γ-space and therefore $\mathbf{L}_{\Sigma f}\check{F}_1 = \mathbf{L}_{\Sigma f}F$ is an ∞-loop space. Let us prove now the first part of the claim:

Consider the diagram that depicts the above constructions for $n = 2$. This diagram is built from the lower right square by taking homotopy fibres:

(B.1.6)
$$\begin{array}{ccccccc} \Omega(\Omega B*\Omega B) & \longrightarrow & X & \longrightarrow & \Omega E*\Omega E & \longrightarrow & \Omega B*\Omega B \\ & & \downarrow & & \downarrow & & \downarrow \\ & & \check{F}_2 & \longrightarrow & E\vee E & \longrightarrow & B\vee B \\ & & \downarrow & & \downarrow & & \downarrow \\ & & F\times F & \longrightarrow & E\times E & \longrightarrow & B\times B \end{array}$$

To facilitate the exposition we prove the equivalence in the the case $n = 2$ the general case follows along the same path. By (1.H.1), in order to prove the claim it is sufficient to show that $\mathbf{L}_{\Sigma f}X \simeq *$. In order to show that we use (1.H.1) again and show that both the fibre and the base in the fibration sequence with X as a total space in (B.1.6) above are killed by $\mathbf{L}_{\Sigma f}$.

To this end it is sufficient to show $\mathbf{L}_{\Sigma f}\Omega(\Omega B*\Omega B) \simeq \mathbf{L}_{\Sigma f}\Omega(\Omega E*\Omega E) \simeq *$. Because then $\mathbf{L}_{\Sigma f}$ certainly kills $(\Omega B*\Omega B)$ and $(\Omega E*\Omega E)$, again by (1.H.1). Now consider e.g. $\mathbf{L}_{\Sigma f}\Omega(\Omega B*\Omega B) \simeq *$. By the usual identities as above:

$$\mathbf{L}_{\Sigma f}\Omega\Sigma(\Omega B\wedge\Omega B) \simeq \Omega\mathbf{L}_{\Sigma^2 f}\Sigma(\Omega B\wedge\Omega B).$$

Therefore it is sufficient to show that $\mathbf{L}_{\Sigma^2 f}\Sigma(\Omega B\wedge\Omega B) \simeq *$. In order to do that we use (1.A.8)(e.10) with $W = \Omega B$ and $n = 0$: We know that B is a 1-connected space since it is killed by assumption by Σf a map of 1-connected spaces. Therefore ΩB is connected.

$\mathbf{L}_{\Sigma f}(\Omega B\wedge\Omega B) \simeq *$ since $\mathbf{L}_f\Omega B \simeq *$ and $\mathbf{P}_{S^1}\Omega B \simeq *$.

thus: $\mathbf{L}_{\Sigma^2 f}\Sigma(\Omega B\wedge\Omega B) \simeq *$. By the same argument we get $\mathbf{L}_{\Sigma^2 f}\Sigma(\Omega E\wedge\Omega E) \simeq *$.

This proves our claim since it implies:

$$\mathbf{L}_{\Sigma f}X \simeq \mathbf{L}_{\Sigma f}(\Omega E*\Omega E) \simeq *.$$

Therefore $\mathbf{L}_{\Sigma f}F$ is an ∞-loop space and, in particular, we can write: $\mathbf{L}_{\Sigma f}F = \Omega Y$.

We now prepare the ground to the use of the key lemma (B.2) below:

Claim: $\mathrm{map}_*(\Sigma^2 F, Y) \simeq \mathrm{map}_*(\Sigma F, \Omega Y) \simeq *$.
This follows from universality (1.A) as follows: We have a factorization:

$$\begin{array}{ccc} \Sigma F & \longrightarrow & \Omega Y = \mathbf{L}_{\Sigma f}F \\ \downarrow & \nearrow_{\exists !} & \\ \mathbf{L}_{\Sigma f}\Sigma F & & \end{array}$$

in which:

Claim: $\mathbf{L}_{\Sigma f}\Sigma F \simeq *$. Moreover: $\mathbf{L}_f F \simeq *$ (1.A.8,e.10).

This is clear from (1.H.1) for the fibration:

$$\Omega B \to F \to E$$

and $\mathbf{L}_f \Omega B \simeq \Omega \mathbf{L}_{\Sigma f} B \simeq \Omega * \simeq *$. All the more so $\mathbf{L}_{\Sigma f}\Sigma^k F \simeq *$, and thus any amp $\Sigma^k F \to \Omega Y$ is null which is equivalent to the claim.

The claim being proven we can conclude the proof of (B.1) just as we did in the proof of (b.1.1) below. from Bousfield's key lemma (B.2) below that $F \to \mathbf{L}_{\Sigma f}F$ factors through the universal GEM associated with F, namely the infinite symmetric product: $SP^\infty F$.

$$\begin{array}{ccc} & & SP^\infty F \\ & \nearrow & \downarrow \\ F & \longrightarrow & \Omega Y = \mathbf{L}_{\Sigma f}F \end{array}$$

But $SP^\infty F$, the Dold–Thom functor on F, is a GEM. Applying $\mathbf{L}_{\Sigma f}$ to the factorization we get that $\mathbf{L}_{\Sigma f}F$ is a retract of a GEM, since $\mathbf{L}_{\Sigma f} = \mathbf{L}_{\Sigma f}\mathbf{L}_{\Sigma f}$. But a retract of a GEM is a GEM.

The naturality of this GEM structure and its compatibility with ∞-loop structure on $\mathbf{L}_{\Sigma f}F$ constructed above follows as in (B.1.6) above from the naturality of construction. This concludes the proof of Theorem B.1. ∎

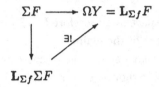

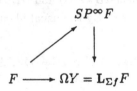

B.2 Bousfield's Key Lemma: *Let X be a connected space, and let Y be a simply-connected space. Assume* $\mathrm{map}_*(\Sigma^2 X, Y) \simeq *$. *Then* $\mathrm{map}_*(X, \Omega Y) \simeq \mathrm{map}_*(SP^k X, \Omega Y)$ *for any* $k \geq 1$ [B-4, 6.9].

B.3 Remark: A proof of (B.2) was given earlier in Chapter 4 using only some basic material about homotopy colimits. A way to understand (2.2) is to interpret it as saying that the space $\Sigma SP^k X$ can be built by successively gluing together copies of $\Sigma^\ell X$ for $\ell \geq 1$ with precisely one copy for $\ell = 1$. Since the higher suspension $\Sigma^{2+j} X$ $(j \geq 0)$ will not contribute anything to $\mathrm{map}_*(\Sigma\, SP^\ell X, Y)$, we are left with $\mathrm{map}_*(\Sigma X, Y)$.

GEM error terms in cofibrations: We conclude this section by presenting two symmetrically looking results. The first is just (B.1.1) above and the second is in fact a dual theorem for the localization of cofibration as its corollary B.4.1 shows.

Corollary B.1.1: *For any X and $f: A \to B$ in $\mathcal{S}$, if* $\mathbf{L}_{\Sigma f} X \simeq *$ *then* $\mathbf{L}_{\Sigma^2 f} X$ *has a natural structure of a GEM.*

The proof was given above.

B.4 Theorem: *For any space X and any map in $\mathcal{S}$, if* $\mathbf{L}_f \Sigma^2 X \simeq *$ *then* $\mathbf{L}_f \Sigma X$ *has a natural structure of a GEM.*

This has a corollary that is a sort of dual to Theorem B.1:

B.4.1 Corollary: *Let* $\Sigma A \to \Sigma X \to \Sigma(X/A)$ *be a suspension of a cofibration sequence. If* $\mathbf{L}_f \Sigma X \simeq \mathbf{L}_f \Sigma(X/A) \simeq *$ *then the localization* $\mathbf{L}_f \Sigma A$ *has a natural structure of a GEM.*

Proof: It follows from (3.D) that the assumption implies $\mathbf{L}_f \Sigma^2 A \simeq *$ using the Puppe sequence. Thus from (B.4) we get the desired result.

Proof of B.4: This follows formally from (B.1.1). First notice that $\mathbf{L}_f \Sigma X \simeq \mathbf{L}_f \mathbf{P}_{\Sigma^2 X} \Sigma X$. This follows immediately from the assumption $\mathbf{L}_f \Sigma^2 X \simeq *$ by (1.D.7). But from (B.1.1) applied to $\mathbf{P}_{\Sigma^2 X}$ we get that $\mathbf{P}_{\Sigma^2 X} \Sigma X$ is a GEM— since $\mathbf{P}_{\Sigma X} \Sigma X \simeq *$. Therefore by (4.B.4) it follows that $\mathbf{L}_f \mathbf{P}_{\Sigma^2} \Sigma X$ is also a GEM, hence the conclusion follows.

The above result has an interesting corollary that will be used later on. This corollary is a partial inverse to elementary fact 1.A.8,(e.10).

B.5 Proposition: *For any map f and space X, if* $\mathbf{L}_{\Sigma f} \Sigma^2 X \simeq *$ *then* $\mathbf{L}_f \Sigma X \simeq *$.

B.6 COROLLARY: *For any spaces X, Y and integer $k \geq 1$ one has: $\Sigma X < \Sigma Y$ iff $\Sigma^k X < \Sigma^k Y$.*

Remark: Of course this is not true if we omit one suspension from the assumption and conclusion, since X may be an acyclic space whose suspension is equivalent to a point.

Proof of B.5: By (B.4) we know that the map $\Sigma X \to \mathbf{L}_f \Sigma X$ is a map of ΣX to a loop space, in fact to an infinite loop space. Therefore it factors, up to homotopy, through the canonical map to the James functor on ΣX: $\Sigma X \to \Omega\Sigma\Sigma X$. Thus it is sufficient to check that localization of the latter $\mathbf{L}_f \Omega\Sigma^2 X$ is contractible. But using (3.A.1) this is exactly our assumption.

Proof of B.6: This follows immediately from (B.5) by taking the map f to be say $\Sigma X \to *$.

C. The nullification applied to fibrations and function complexes

In this section we consider the action of nullification functor $\mathbf{P}_A$ on fibration sequences and on function complexes. The statements of the main theorems appear in Section A above. Then in Section D below we consider general localizations. In all cases one has to assume that the maps with respect to which one localizes are at least one suspension maps, or equivalently, the spaces and maps have a loop structure.

We employ repeatedly the following diagram:

(C.1)
$$\begin{array}{ccccccc}
Y & \longrightarrow & \overline{X} & \longrightarrow & \overline{\mathbf{L}}E & \longrightarrow & \overline{\mathbf{L}}B \\
\downarrow & & \downarrow & & \downarrow & & \downarrow \\
\Omega B & \longrightarrow & F & \longrightarrow & E & \longrightarrow & B \\
\downarrow & & \downarrow & & \downarrow & & \downarrow \\
\Omega \mathbf{L}B & \longrightarrow & X & \longrightarrow & \mathbf{L}E & \longrightarrow & \mathbf{L}B
\end{array}$$

where $\mathbf{L}$ is $\mathbf{L}_f$, a localization functor for some map f, and $\overline{\mathbf{L}}(-)$ denotes the homotopy fibre of the coaugmentation. Let X, $\overline{X}$, Y be the appropriate homotopy fibres that render every sequence of two collinear arrows a fibre sequence. The natural map $F \to X$ is the map $F \to$ fibre $(\mathbf{L}E \to \mathbf{L}B)$. We get a well defined homotopy class $\mathbf{L}F \to X$ or $\mathbf{L}F \to$ fibre $(\mathbf{L}E \to \mathbf{L}B)$, as in (A.3) above. Now let us restrict to the case $\mathbf{L} = \mathbf{P}_A$. In that case the fibration sequence $\overline{X} \to F \to X$ is preserved by $\mathbf{L} = \mathbf{P}_A$ by (3.D.3)(2) above, since X is A-null by (1.A.8) (e.3) above. Therefore, upon taking $\mathbf{P}_A$ of that fibration sequence, we get a fibration

$$\mathbf{P}_A \overline{X} \to \mathbf{P}_A F \to \mathbf{P}_A X = X$$

or

$$\mathbf{P}_A\overline{X} \to \mathbf{P}_A F \to \text{ fibre } (\mathbf{P}_A E \to \mathbf{P}_A B).$$

Therefore $\mathbf{P}_A\overline{X}$ is the 'error-term' of $\mathbf{P}_A$, and whenever $\mathbf{P}_A\overline{X}$ is a (poly)GEM, $\mathbf{P}_A$ preserves the fibration $F \to E \to B$ up to a (poly)GEM (A.1 above).

C.2 OBSERVATION: *If a localization functor* $\mathbf{L} = \mathbf{P}_A$ *commutes up to homotopy with* ΩB, *i.e. if* $\mathbf{L}\Omega B \xrightarrow{\simeq} \Omega\mathbf{L}B$, *then it will preserve any fibration over* B. *Moreover, for a general localization functor* $\mathbf{L} = \mathbf{L}_f$, *if we assume in addition* $\mathbf{L}_f\Omega E \simeq \Omega\mathbf{L}_f E$ *and* $\mathbf{L}_f\Omega F \simeq \Omega\mathbf{L}_f F$, *then we can conclude that* $\mathbf{L}_f$ *preserves the fibration* $F \to E \to B$.

Proof: We use diagram (C.1). Under the assumption the map $\Omega B \to \Omega\mathbf{L}B$ is a $\mathbf{P}_A$-localization map and thus, by (1.H.2) above, implies $\mathbf{L}Y \simeq *$ for $\mathbf{L} = \mathbf{P}_A$. Therefore $\mathbf{L}\overline{X} \simeq \mathbf{L}\overline{\mathbf{L}}E$ by (1.H.1) above. Since $\mathbf{L}(\overline{\mathbf{L}}E) \simeq *$ (1.H.2) we get $\mathbf{L}\overline{X} \simeq *$. Since the fibration $\overline{X} \to F \to X$ is preserved in such a case, $\mathbf{L}F = \mathbf{L}X \simeq X$; and since X is local, being the fibre of local spaces we are done: $\mathbf{L}F \simeq \mathbf{L}X \simeq X$. The proof in the general case is similar.

PROOF OF (A.5) We argue with diagram (C.1) using Theorem 3.A.1. We now read (C.1) with $\mathbf{L} = \mathbf{P}_A, \overline{\mathbf{L}} = \overline{\mathbf{P}}_A$. By (1.H.1) above we get $\mathbf{P}_A\overline{\mathbf{P}}_A B \simeq \mathbf{P}_A\overline{\mathbf{P}}_A E \simeq *$. Therefore $(A = \Sigma A')$ we can use Theorem B.1 to deduce $\mathbf{P}_A\overline{X} \simeq$ GEM.

Notice that $\overline{X} \to F \to X$ in diagram (C.1) is a fibration with an A-null base space X, so by (3.D.3)(2) it is preserved under $\mathbf{P}_A$. Thus the fibre of $\mathbf{P}_A F \to X = \mathbf{P}_A X$ is $\mathbf{P}_A\overline{X}$, an A-null GEM. Finally, $\mathbf{P}_{A'}\overline{X} \simeq *$ since $\mathbf{P}_{A'}$ kills both Y and $\overline{\mathbf{P}}_A E$.

The following is a generalization of Theorem B.1 and a weak inverse to (1.H.1) that follows easily by chasing the diagram (C.1) above.

C.3 COROLLARY: *In any fibration* $F \to E \to B$ *that induces an equivalence* $\mathbf{P}_{\Sigma A}E \xrightarrow{\simeq} \mathbf{P}_{\Sigma A}B$, *the nullification* $\mathbf{P}_{\Sigma A}F$ *is a GEM.*

C.4 Localization of function complexes

We now show that the above theory gives rather good control of the localization of function complexes when we consider nullification with respect to finite-dimensional suspension spaces. The main result here is the following strong version of [B-4, 8.3].

C.5 THEOREM: *If* $V, W \in \mathcal{S}_*$ *are connected spaces with* W *finite dimensional, then the homotopy fibre* $F = F_\varphi$ *over any vertex* $\varphi : V \to \mathbf{P}_{\Sigma W}X$ *of the natural map*

$$\mathbf{P}_{\Sigma W}(X^V) \to (\mathbf{P}_{\Sigma W}X)^V$$

has vanishing homotopy groups $\pi_i F \simeq 0$ *for* $i \geq \dim W + 1 = \dim \Sigma W$.

The proof of this theorem is based on several useful lemmas and is given after (C.8) below.

C.6 LEMMA: *If, for some integer* $n \geq \dim W$, *the abelian group* G *satisfies* $\mathbf{P}_W K(G,n) \simeq K(G,n)$ *and* $\mathbf{P}_W K(G,n+1) \simeq *$, *then* $G \simeq 0$.

Another formulation of the same lemma is

C.7 LEMMA: *If* $\mathbf{P}_W K(G,n+1) \simeq *$ *with* $n \geq \dim W$, *then* $\mathbf{P}_W K(G,n) \simeq *$.

C.7.1 REMARK: Clearly the opposite direction is true without any assumption on $\dim W$ or n. Intuitively we think about an equation $\mathbf{P}_W K(G,n+1) \simeq *$ or $W < K(G,n+1)$ as saying that $\mathbb{Z}W \simeq \prod_{i=1}^{\dim W} K(H_i W, i)$ has groups in dimensions not above $n+1$ that 'support' the group G, i.e. from which this group can be built, so that killing these $H_i(W)$ kills the group G too. In that case, lowering n by one without coming below the dimension of W will leave the inequality $W < K(G,n+1)$ true in a stronger form: $W < K(G,n)$.

In other words, the inequality:

$$\prod_{i=1}^{m} K(G_i, i) < K(G,m)$$

is equivalent to:

$$K(\bigoplus_i G_i, m) < K(G,m).$$

Proof of Lemma C.6: We are given that $K(G,n)$ is W-null. This can be written as $\Omega\mathrm{map}_*(W, K(G,n+1)) \simeq *$. Since we assume $n \geq \dim W$, this function complex is connected before taking the loop and so its loop, being contractible, implies that the function complex itself is $\simeq *$. Thus $K(G,n+1)$ is also W-null. But $\mathbf{P}_W K(G,n+1) \simeq *$ by assumption, so $G \simeq 0$.

Proof of Lemma C.7: Let $K(\overline{G},\ell)$ be a factor in the GEM space $\mathbf{P}_W K(G,n)$ and we wish to show $\overline{G} \simeq 0$. To this end we use (C.6) that we have just proven. In order to satisfy the conditions of (C.6) we need to verify three things:

(1) The integer $\ell \geq \dim W$. This is clear since by assumption $n \geq \dim$ W and we know from (4.B.2) that $\ell \geq n$.

(2) $\mathbf{P}_W K(\overline{G},\ell) = K(\overline{G},\ell)$. This follows since by definition the latter is a retract of a W-null space.

(3) $\mathbf{P}_W K(\overline{G},\ell 1) \simeq *$. This is true since the map $K(G,n) = \Omega K(G,n+1) \to \mathbf{P}_W \Omega K(G,n+1)$ is a loop map by (3.A) above and therefore, by induction on the connectivity of range, we may assume that $K(G,n) \to K(\overline{G},\ell)$ is also a loop map. Taking the classifying space functor on the latter map we get $K(G,n+1) \to$

$K(\overline{G}, \ell+1)$. But by Lemma (C.8) below $K(\overline{G}, \ell+1)$ is a factor in $\mathbf{P}_W K(G, n+1) \simeq *$, thus $\overline{G} \simeq 0$ as needed.

C.8 LEMMA: (Compare [B-4, 4.7]) *Let Y, Y' be connected spaces. If the map $\Omega g : \Omega Y \to \Omega Y'$ is a loop map with the induced map $\mathbf{P}_W \Omega g$ being a weak equivalence, then so is the map $\mathbf{P}_W g : \mathbf{P}_W Y \to \mathbf{P}_W Y'$.*

Proof: This follows at once from the formula for localization of loop spaces (3.A.1) above.

Before proving the theorem (C.5) above we deduce the following neat consequence from the lemma above.

C.9 THEOREM: *If $\Sigma X < Y$ and $\dim \Sigma X \leq \mathrm{conn} Y$, then* (i) $\Sigma^2 X < Y$ *and* (ii) $\Sigma X \ll Y$.

Proof: Since $\mathbf{P}_{\Sigma X} Y \simeq *$, we get from GEM theory (B.1.1) above that $J \equiv \mathbf{P}_{\Sigma^2 X} Y$ is a GEM. The space J is $\Sigma^2 X$-null and $\mathbf{P}_{\Sigma X} J \simeq *$. Let $K(G, n)$ be a factor in J. We first show that $n \leq \mathrm{conn} Y$. Assume $n > \mathrm{conn} Y \geq \dim \Sigma X$. Since $\mathbf{P}_{\Sigma X} K(G, n) \simeq *$ one gets $\mathbf{P}_{\Sigma^2 X} K(G, n+1) \simeq *$. But $K(G, n)$ is $\Sigma^2 X$-null and $n \geq \dim \Sigma^2 X$, so we are under the assumption of Lemma C.6 with $W = \Sigma^2 X$. Therefore $G \simeq 0$. Hence each factor $K(G, n)$ with $G \neq 0$ in J satisfies $n \leq \mathrm{conn} Y$. But then $[Y, K(G, n)] \simeq *$, thus the map $Y \to J = \mathbf{P}_{\Sigma^2 X} Y$ is null homotopic and, by idempotency $\mathbf{P}_A$, we get $\mathbf{P}_{\Sigma^2 X} Y \simeq *$ as claimed. From (3.B.3) above we get (ii).

We now turn to

Proof of Theorem C.5: We proceed by induction on the skeleton of V. For $V = S^1$ we must consider the homotopy fibre F of the map

$$\mathbf{P}_{\Sigma W} \Omega X \to \Omega \mathbf{P}_{\Sigma W} X$$

over a component of the base. We have shown in (A.5) that this fibre is a ΣW-null GEM with $\mathbf{P}_W F \simeq *$. Consider a typical factor $K(G, n)$ in the GEM F. This factor is again ΣW-null and, by Lemma C.7 above, $\mathbf{P}_{\Sigma W} K(G, n+1) \simeq *$. Therefore by Lemma C.6 unless $n < \dim \Sigma W$ one has $G \simeq 0$. It follows that $\pi_i F \simeq 0$ for all $i \geq \dim \Sigma W$, as claimed.

To continue the proof of the theorem we first notice that, by induction, the claim of the theorem is true for $V = S^n$ by examining the homotopy fibres of the maps in a square of the form (for $n = 2$):

$$\begin{array}{ccc} \mathbf{P}_{\Sigma W} \Omega^2 X & \longrightarrow & \Omega \mathbf{P}_{\Sigma W} \Omega X \\ c \downarrow & & \downarrow \\ \Omega^2 \mathbf{P}_{\Sigma W} X & = & \Omega^2 \mathbf{P}_{\Sigma W} X \end{array}$$

It follows, by considering corresponding exact sequences of fibrations in the square, that the fibre of the map c has vanishing homotopy groups above dimension $\dim W + 1$ for this particular $V = S^n$.

Now for a general V we first use a similar square-argument for the fibration induced by the cofibration $S^k \to V^1 \to V$, where we assume that S^k and V^1, playing the role of V as exponents in (C.5), already satisfy the conclusion of Theorem C.5.

This gives by finite induction the claim for any finite V. But since maps from an infinite complex V are just the homotopy (inverse) limit of the maps from the finite-dimensional skeleta, and since taking homotopy fibres commutes with taking the linear direct limit (Appendix HL), we get the vanishing of the higher homotopy groups as needed for any V. This completes the proof. ∎

D. Localization with respect to a double suspension map

In this section we apply the above material to a discussion of the fibre of $\mathbf{L}_f\Omega Y \to \Omega\mathbf{L}_f Y$ and prove Theorem A.6 for a general map $f : A \to D$ and a double loop space $Y = \Omega^2 X$. The main observation is that the difference between $\mathbf{L}_f$ and $\mathbf{L}_{C(f)}$ (where $C(f) = D \cup CA =$ the mapping cone of f) is an f-local GEM.

First we make the following simple observation about any cofibration:

D.1 PROPOSITION: *If $A \xrightarrow{f} D \xrightarrow{h} C$ is a cofibration with cofibre $C = D \cup CA$, then any f-local space is C-null and any A-null space is h-local.*

Proof: This is immediate from the definition and the fact that, for any space X, the sequence:

$$\mathrm{map}_*(C, X) \xrightarrow{\overline{h}} \mathrm{map}_*(D, X) \xrightarrow{\overline{f}} \mathrm{map}_*(A, X)$$

is a fibre sequence with $\mathrm{map}_*(C, X)$, the homotopy fibre *over the null component.* So if $\mathrm{map}_*(A, X) \simeq *$ then $\overline{h}$ is an equivalence, while if $\overline{f}$ is an equivalence then $\mathrm{map}_*(C, X) \simeq *$.

D.2 REMARK: So when we consider a Barratt–Puppe sequence:

$$A \xrightarrow{f} D \xrightarrow{h} C \to \Sigma A \xrightarrow{\Sigma f} \Sigma D \to \Sigma C \to \Sigma^2 A \to \Sigma^2 D \to \cdots$$

as we pick maps and spaces more and more to the right: $A, h, \Sigma A, \Sigma h \cdots$, being local or periodic or null with respect to these maps and spaces, becomes a strictly weaker condition.

D.3 COROLLARY: *As a direct result of (D.1) we get $\mathbf{P}_A\mathbf{L}_h = \mathbf{P}_A$ and $\mathbf{L}_f\mathbf{P}_C = \mathbf{L}_f$.*

We are now ready to prove the very useful

D.4 LEMMA: *For any map $f : A \to D$ the fibre of $\mathbf{L}_{\Sigma^2 f}X \to \mathbf{P}_{\Sigma C}X$ is a $\Sigma^2 f$-null GEM that localizes to a point under ΣC, where C is $C(f)$, the mapping cone of f.*

Proof: The cofibration sequence depicted above together with (D.1) show that any ΣC periodic space is $\Sigma^2 f$-local. Therefore (D.3):

$$\mathbf{P}_{\Sigma C}\mathbf{L}_{\Sigma^2 f} \simeq \mathbf{P}_{\Sigma C}.$$

Hence, the following map is the ΣC-localization:

$$\mathbf{L}_{\Sigma^2 f}X \to \mathbf{P}_{\Sigma C}X,$$

and (1.H.1) its fibre F satisfies $\mathbf{P}_{\Sigma C}F \simeq *$. Moreover, F is $\Sigma^2 C$ null, being $\Sigma^2 f$-local. But by Corollary B.1.1 above we get that $\mathbf{P}_{\Sigma^2 C}F$ is a GEM. So $F = \mathbf{P}_{\Sigma^2 C}F$ is a GEM which is $\Sigma^2 f$-null.

This can be rewritten in a different form:

D.5 COROLLARY: *There is a natural map for any f:*

$$\sigma : \mathbf{L}_f\Omega^2 X \to \Omega\mathbf{P}_{C(f)}\Omega X,$$

whose fibre is an f-local GEM.

Proof: Consider the map

$$\mathbf{L}_{\Sigma^2 f}X \to \mathbf{P}_{\Sigma C}X$$

given in the proof of the lemma above. Since (3.A.1) $\mathbf{L}_f\Omega Y \simeq \Omega\mathbf{L}_{\Sigma f}Y$, we can rewrite the map as

$$\overline{W}^2\mathbf{L}_f\Omega^2 X \to \overline{W}\mathbf{P}_C\Omega X$$

with the fibre being $\Sigma^2 C$-local GEM.

Looping it down twice we get the desired map $\mathbf{L}_f\Omega^2 X \to \Omega\mathbf{P}_C\Omega X$ with $C = C(f)$, the mapping cone of f. But now the fibre is a f-local GEM, being the double loop of a $\Sigma^2 f$-null space.

Remark: The usefulness of (D.4) and (D.5) follows from the fact that $\mathbf{P}_W$, the nullification functor, behaves much more nicely with respect to fibrations as in Section C and (A.5) than $\mathbf{L}_f$ for a general map f.

We now turn to the Main Theorem about preservation of fibration under $\mathbf{L}_{\Sigma^2 g}$:

D.6 THEOREM: *Let $g : A \to D$ be a map of connected spaces, and $F \to E \xrightarrow{p} B$ be a fibration sequence with B connected. Then the homotopy fibre Δ of the natural map*

$$\mathbf{L}F \to \text{fibre}(\mathbf{L}E \to \mathbf{L}B),$$

where $\mathbf{L} = \mathbf{L}_{\Sigma^2 g}$, satisfies

(1) *Δ is a polyGEM and $\Omega\Delta$ is an oriented polyGEM,*
(2) *Δ is $\Sigma^2 g$-local,*
(3) *$\mathbf{L}_g \Delta \simeq *$.*

Proof: The proof is presented after the next corollary.

Remark: Conditions (1), (2) and (3) force the error term Δ to be 'small' at each prime p, see Chapter 6 below. Here is one useful corollary which may hold under a weaker assumption.

D.6.1 COROLLARY *If Y is a polyGEM, then for any $g : A \to B$ the localization of the double loop space $\mathbf{L}_g \Omega^2 Y$ is again a polyGEM.*

Proof: By (4.B.4) $\mathbf{L}_g$ (GEM) is again a GEM. We now proceed by induction: Let $W \to X \to Y$ be a fibration with $\mathbf{L}_g \Omega^2 X, \mathbf{L}_g \Omega^2 Y$ polyGEMs. Now by the above theorem (D.6) using adjunction $\Omega^2 \mathbf{L}_{\Sigma^2 g} = \mathbf{L}_g \Omega^2$, we get a sequence of fibrations where the classification map to $\Omega\Delta$ is obtained using (3.A.1) above:

$$\Omega^2\Delta \to \mathbf{L}_g\Omega^2 W \to \text{ Fibre } (\mathbf{L}_g\Omega^2 X \to \mathbf{L}_g\Omega^2 Y) \to \Omega\Delta,$$

where Δ is as in (D.6). By (D.6)(1), $\Omega\Delta$ is a polyGEM and so, by the inductive definition (A.2), $\mathbf{L}_g\Omega^2 W$ is a polyGEM as needed.

Proof of (D.6):
By (1.A.8) (e.3), (2) is immediate. We first deal with (3) by:

D.7 LEMMA: $\mathbf{L}_g(\Delta) \simeq *$.

Proof: In diagram (C.1) with $\mathbf{L} = \mathbf{L}_{\Sigma^2 g}$ we get, in view of (3.A.1) above, that Y is in fact the homotopy fibre of the localization map fibre of $(\Omega B \to \mathbf{L}_{\Sigma g}\Omega B)$ and therefore, by (1.H.3) above, $\mathbf{L}_g Y \simeq *$. So by (1.H.1) and (1.H.2) above we get $\mathbf{L}_g \overline{X} \simeq \mathbf{L}_g(\overline{\mathbf{L}E}) \simeq *$. Now consider the diagram of fibrations below which is derived from the relevant column in (C.1), turned into the middle row here, and Δ is *defined* as the homotopy fibre in the bottom row. Notice that X is $\Sigma^2 g$-local, being a fibre of a map between such spaces.

$$\begin{array}{ccccc}
\overline{\mathbf{L}}_{\Sigma^2 g}F & = & \overline{\mathbf{L}}_{\Sigma^2 g}F & \longrightarrow & * \\
\downarrow & & \downarrow & & \downarrow \\
\overline{X} & \longrightarrow & F & \longrightarrow & X \\
\downarrow & & \downarrow & & \downarrow \\
\Delta & \longrightarrow & \mathbf{L}_{\Sigma^2 g}F & \longrightarrow & X = \mathbf{L}_{\Sigma^2 g}X
\end{array}$$

By (3.A) again one gets $\mathbf{L}_g\overline{\mathbf{L}}_{\Sigma^2 g}F \simeq *$, and therefore, as usual, $\mathbf{L}_g\Delta \simeq \mathbf{L}_g\overline{X} \simeq *$ as needed.

D.8 Remark: Therefore by the Main Theorem B.1, we get $\mathbf{L}_{\Sigma g}(\Delta)$ is a GEM. However, Δ itself is not an Σg-local space, so we cannot conclude in general for $\mathbf{L}_g$ that Δ is a GEM even for $\mathbf{L} = \mathbf{L}_{\Sigma^3 g'}$. To circumvent this difficulty we use (D.4) above and must relax GEM to polyGEM.

Proof of D.6(1):

Proceeding with the proof of D.6 we compare $\mathbf{P}_{\Sigma C(g)}$ with $\mathbf{L}_{\Sigma^2 g}$, as follows using Lemma D.4 above.

The diagram is obtained in the usual way by backing up from the lower right square in which the vertical arrows are W-localization maps in light of (D.3) above.

$$\text{(D.9)} \qquad \begin{array}{ccccc}
F_3 & \longrightarrow & (GEM)_1 & \longrightarrow & (GEM)_2 \\
\downarrow & & \downarrow & & \downarrow \\
F_2 & \longrightarrow & \mathbf{L}_{\Sigma^2 g}E & \longrightarrow & \mathbf{L}_{\Sigma^2 g}B \\
\downarrow & & \downarrow & & \downarrow \\
F_1 & \longrightarrow & \mathbf{P}_{\Sigma C(g)}E & \longrightarrow & \mathbf{P}_{\Sigma C(g)}B
\end{array}$$

As a result of (D.4) above, the homotopy fibres denoted by $(\mathrm{GEM})_1$, $(\mathrm{GEM})_2$ are in fact GEM spaces.

It follows by definition that F_3 is a polyGEM, being the homotopy fibre of maps between GEMs. On the other hand, using Theorem A.5 to compare $\mathbf{P}_W$ (fibre) and fibre $\mathbf{P}_W(E \to B)$, we get the following diagram in which the central vertical sequence measures the difference between the fibre of the localization and the localization of the fibre. The spaces F_i are from (D.9) above.

(D.10)
$$\begin{array}{ccccc}
(\text{PolyGEM})_2 & \longrightarrow & \Delta & \longrightarrow & (\text{GEM})_4 \\
\downarrow & & \downarrow & & \downarrow \\
(\text{GEM})_3 & \longrightarrow & \mathbf{L}_f F & \longrightarrow & \mathbf{P}_{\Sigma C(g)} F \\
\downarrow & & \downarrow & & \downarrow \\
(\text{PolyGEM}) = F_3 & \longrightarrow & F_2 & \longrightarrow & F_1 = \text{Fib}(\mathbf{P}_{\Sigma C(g)}(E \to B))
\end{array}$$

Notice again that diagram (D.10) is obtained by backing up from the lower right square taking homotopy fibres. The vertical arrows in that square are this time comparison arrows from the localization of the fibre to the fibre of localizations constructed just as in the proof of Theorem A.4 above according to the demands of Definition A.1 above.

The above diagram shows that F_3 is a polyGEM by Theorem A.5 as we saw in (D.9), and $(\text{GEM})_4$ is a GEM since F_1 is given as a fibre of $\mathbf{P}_W(E \to B)$. By (D.4) we get that $(\text{GEM})_3$ is also a GEM. Hence by definition Δ is a polyGEM. Backing up the fibration in the top row we also conclude therefore that $\Omega\Delta$ is an oriented polyGEM by (A.2), as needed in (D.6)(1). This completes the proof of (D.6). ∎

E. The functor $\mathbf{CW}_A$ and fibrations

In this section we consider the behaviour of fibration sequences under $\mathbf{CW}_A$. The results of Chapter 3 above suggest that we can expect here GEM error terms just as in applying $\mathbf{P}_A$ to such sequences for a suspension space A. Unfortunately we get somewhat weaker results. For example, the analogue to the Main Theorem above is that, for a connected A, $\mathbf{CW}_{\Sigma A} X \simeq *$ implies that $\mathbf{CW}_{\Sigma A} X$ is a GEM. However, we can only show that the fibre of $\mathbf{CW}_{\Sigma^2 A} X \to \mathbf{CW}_{\Sigma A} X$ is a polyGEM.

The slogan is: 'Whenever the function complex $\text{map}_*(\Sigma A, X) \simeq S$ is homotopically discrete, the space $\mathbf{CW}_{\Sigma A} X$ is a GEM thus the above function <u>set</u> S has a natural <u>abelian</u> group structure' (compare [B-4]).

Recall that if $\mathbf{P}_{\Sigma A} X \simeq *$, then $\mathbf{P}_{\Sigma^2 A} X$ is a GEM. Here we have a similar result about $\mathbf{CW}_A X$.

E.1 Theorem: *Assume* $\mathbf{CW}_{\Sigma^2 A} X \simeq *$. *Then* $\mathbf{CW}_{\Sigma A} X$ *has a natural structure of a GEM.*

Proof: Consider the natural square of maps associated to any space X:

$$\begin{array}{ccc}
j\colon \mathbf{CW}_{\Sigma A} X & \longrightarrow & X \\
\downarrow & & \downarrow \simeq \\
\mathbf{P}_{\Sigma^2 A} j\colon \mathbf{P}_{\Sigma^2 A} \mathbf{CW}_{\Sigma A} X & \longrightarrow & \mathbf{P}_{\Sigma^2 A} X
\end{array}$$

Our assumption is equivalent to $\operatorname{map}_*(\Sigma^2 A, X) \simeq *$, i.e. X is $\Sigma^2 A$-null and the right vertical map is an equivalence. Now since $\mathbf{P}_{\Sigma A}\mathbf{CW}_{\Sigma A}X \simeq *$ for any X, A above, we get from (B.1.1) that $\mathbf{P}_{\Sigma^2 A}\mathbf{CW}_{\Sigma A}X$ is a GEM. Therefore we conclude that the canonical map $\mathbf{CW}_{\Sigma A}X \to X$ factors up to homotopy through a GEM. Since by Lemma 4.B.3 $\mathbf{CW}_A$ (GEM) is always a GEM, we conclude that $\mathbf{CW}_{\Sigma A}X$ is a retract of a GEM, thus a GEM.

E.2 PROBLEM: Is it true that for any A and a polyGEM X the space $\mathbf{CW}_A X$ is also a polyGEM?

The following may not be the best possible result:

E.3 THEOREM: *For any $A, X \in \mathcal{S}_*$ the homotopy fibre F of*

$$j\colon \mathbf{CW}_{\Sigma^2 A}X \to \mathbf{CW}_{\Sigma A}X$$

is a (twisted) polyGEM.

Proof: This follows from (B.1). Notice that $\mathbf{P}_{\Sigma A}$ kills both the domain and range of j above. Therefore $\mathbf{P}_{\Sigma A}F$ is a GEM. But $\operatorname{map}_*(\Sigma^2 A, j)$ is a homotopy equivalence. Therefore $\operatorname{map}_*(\Sigma^2 A, F) \simeq *$ and F is $\Sigma^2 A$-null. Therefore by (B.1.2) the fibre of the map from F to $\mathbf{P}_{\Sigma A}F$ is also a GEM, thus F itself sitting in a fibration between two GEMs is a polyGEM and we are done. ∎

E.4 COMMUTING $\mathbf{CW}_A$ WITH TAKING HOMOTOPY FIBRES: We will now address the question of the preservation of fibration by $\mathbf{CW}_A$. Looking at $A = S^n$ we see immediately that $X\langle n\rangle = \mathbf{CW}_{S^{n+1}}$ is the n-connected cover of X and so it does not preserve fibration in general. But again in this example fibrations are nearly preserved up to a single Eilenberg–Mac Lane space. In general we shall see that when A is a suspension the functor $\mathbf{CW}_A$ 'almost' preserves fibrations, the error term being a GEM or a polyGEM.

In order to measure the extent to which $\mathbf{CW}_A$ preserves fibrations we will now compare the fibre of the CW-approximation with the CW-approximation of the fibre via the following natural map:

(E.5) $\lambda\colon \mathbf{CW}_A F \to \operatorname{Fib}(\mathbf{CW}_A E \to \mathbf{CW}_A B)$

This is associated to any fibration sequence $F \to E \xrightarrow{p} B$ over a connected B. For $E \simeq *$ we get, as a special case, a map $\mathbf{CW}_A \Omega B \to \Omega\mathbf{CW}_A X$ for any space B.

In order to construct λ one notices that the fibre of the map $\mathbf{CW}_A(p)$, denoted here by $\operatorname{Fib}\mathbf{CW}_A(p)$, maps naturally to F. This map induces an equivalence on function complexes $\mathrm{map}_*(A, \mathrm{Fib}(\mathbf{CW}_A p)) \to \mathrm{map}_*(A, F)$ since $\mathrm{map}_*(A, -)$ commutes with taking homotopy fibres. Therefore, by the universal property (2.E.8) there is a factorization $\mathbf{CW}_A F \to \mathrm{fib}(\mathbf{CW}_A(p))$ unique up to homotopy.

Now in general one shows:

E.6 PROPOSITION: *Whenever $A = \Sigma^2 A'$ is a double suspension, the homotopy fibre Δ of the above natural λ is an extension of two GEM spaces, i.e. it is a (twisted) 2-polyGEM:*

$$(GEM)_2 \to \Delta \to (GEM)_1.$$

Moreover, Δ is an A-null, A'-cellular space.

Proof: First we notice that by a straightforward argument one shows that $\mathrm{map}_*(A, \Delta) \simeq *$, i.e. Δ is A-null. This is because the map $\mathrm{map}_*(A, \lambda)$ is a homotopy equivalence, since the fibre J of $\Delta = \mathbf{P}_{\Sigma^2 A'}\Delta \to \mathbf{P}_{\Sigma A'}\Delta$ is a GEM (by 5.A.5). By (3.A.2) above both the domain and range of λ are $\Sigma A'$-cellular and thus both are killed by $\mathbf{P}_{\Sigma A'}$. Therefore the condition of (B.1) is satisfied and $\mathbf{P}_{\Sigma A'}\Delta$ is a GEM. This completes the proof. ∎

E.7 COROLLARY: *For $A = \Sigma^2 A'$ the fibre of $\mathbf{CW}_A \Omega X \to \Omega \mathbf{CW}_A X$ is a 2-polyGEM.*

Proof: Apply the above to $\Omega X \to * \to X$.

F. Applications: A generalized Serre theorem, Neisendorfer theorem

It is well known that finite, 1-connected and non-contractible **CW**-complexes have non-trivial homotopy groups in infinitely many dimensions. This has been generalized in many directions, relaxing the assumption of finiteness. In this section we consider a different direction of generalizing. Instead of considering $[S^n, X]$ we will consider $[\Sigma^n A, X]$ for an arbitrary space A: Instead of assuming X is a finite simply connected **CW**-complex we assume X is a *finite ΣA-cellular space for any connected A.* Namely X (2.E.7) a space obtained by a finite number of steps starting with a finite wedge of copies of ΣA and adding cones along maps from ΣA to the earlier step:

$$X \simeq (\bigvee \Sigma A) \cup C\Sigma^{\ell_1} A \cup C\Sigma^{\ell_2} A \cup \cdots \cup C\Sigma^{\ell_k} A \qquad (\ell_i \geq 1).$$

F.1 THEOREM: *Let A be any pointed, connected space of finite type. Let X be any finite ΣA-cellular space, with $\tilde{H}^*(X, \mathbb{Z}/p\mathbb{Z}) \neq 0$ for some p. Then $\pi_i(X, A) = [\Sigma^i A, X] \neq 0$ for infinitely many dimensions $i \geq 0$.*

One immediate corollary is for $X = \Sigma A$.

F.2 COROLLARY: *Let A be any pointed, connected space of finite type with $\tilde{H}_*(A, \mathbb{Z}/p\mathbb{Z}) \not\cong 0$ for some p. There are infinitely many ℓ's for which $[\Sigma^\ell A, \Sigma A] \neq *$.*

Proof: First we note that, since we consider spaces built from ΣA, by a finite number of cofibration steps we get a space which is conic in the sense of [HFLT], namely it is derived from a single point by a finite number of steps taking mapping cones. Now since $\tilde{H}^*(X, \mathbb{F}_p) \neq 0$ for some p, X satisfies the hypothesis of [HFLT]. Their argument now shows that X cannot have a finite generalized Postnikov decomposition, i.e. in the present terminology X cannot be an oriented polyGEM, since X is of finite type.

On the other hand, suppose $[\Sigma^i A, X] \simeq 0$ for $i \geq N$. Then $\text{map}_*(\Sigma^N A, X) \simeq *$ since all the homotopy groups of this space vanish. In other words, X is $\Sigma^N A$-null or $\mathbf{P}_{\Sigma^N A} X \simeq X$. We claim that $\mathbf{P}_{\Sigma A} X \simeq *$. This is true since, by assumption, X is a ΣA-cellular space (Example 2.D.2 above).

But by induction from (B.1.1) we show below (F.6), (F.7) that the homotopy fibre F_N of $\mathbf{P}_{\Sigma^N A} X \to \mathbf{P}_{\Sigma A} X$ is an oriented polyGEM for any connected A, X. However, we just saw that the homotopy fibre of that map is X itself.

But this space by [HFLT] cannot be an *oriented* polyGEM (A.1). This contradiction implies $[\Sigma^i A, X] \not\cong 0$ for infinitely many i's, as needed.

Remark: Notice that in order to prove Corollary F.2 above we do not have to use the heavy result of [HFLT], since $H_* \Omega \Sigma Y$ is a tensor algebra and is not nilpotent. Therefore, already by Moore–Smith [M-S] ΣA cannot be an *oriented* polyGEM (A.1) if $\tilde{H}_*(\Sigma A, \mathbb{Z}/p\mathbb{Z}) \neq 0$. Hence there must be infinitely many maps $\Sigma^\ell A \to \Sigma A$ for any such A.

F.3 THE MAP σ_N: $\mathbf{P}_{\Sigma^N W} X \longrightarrow \mathbf{P}_{\Sigma W} X$

Let X, W be any connected spaces. We now apply (A.8) to analyze the fibre of the map σ_N: $\mathbf{P}_{\Sigma^N W} X \longrightarrow \mathbf{P}_{\Sigma W} X$ Since it follows directly form the definitions that $\mathbf{P}_{\Sigma W} \mathbf{P}_{\Sigma^N W} X \simeq \mathbf{P}_{\Sigma W} X$ the following map is up to homotopy a ΣW-nullification map: σ_N: $\mathbf{P}_{\Sigma^N W} X \longrightarrow \mathbf{P}_{\Sigma W} X$ for any spaces W, X. Consider this map for $N = 2$.

F.4 PROPOSITION: *The homotopy fibre $\mathbb{F}$ of the map σ_2 is a GEM.*

Proof: We know from (1.H.2) that $\mathbf{P}_{\Sigma W} F \simeq *$ since $\mathbb{F}$ is the fibre of ΣW-nullification. Therefore by (B.1.1) we get that $\mathbf{P}_{\Sigma^2 W} F$ is a GEM. However being the homotopy fibre of a map of $\Sigma^2 W$-null spaces $\mathbb{F}$ is also $\Sigma^2 W$-null. Therefore $\mathbb{F}$ is equivalent to its $\Sigma^2 W$-nullification and thus it is a GEM.

The GEM appearing in (F.4) has several special properties that we have already seen in (A.8) and (A.9) above. Therefore one can deduce directly from the argument of (F.4) above:

F.4.1 COROLLARY: *The homotopy fibre $\mathbb{F}$ of the map σ_2 is a GEM whose space of pointed self maps* $\mathrm{map}_*(\mathbb{F}, \mathbb{F})$ *is discrete.*

Proof: See (A.8)–(A.10) above.

It is killed by $\mathbf{P}_{\Sigma W}$ and it is also $\Sigma^2 W$-null. We now proceed to consider the fibre of σ_N. We use the usual comparison diagram:

(F.4.2)
$$\begin{array}{ccccc}
\mathbb{F}_2 & \longrightarrow & \mathbb{F}_3 & \longrightarrow & \mathbb{F}_1 \\
\| \downarrow & & \downarrow & & \downarrow \\
\mathbb{F}_2 & \longrightarrow & \mathbf{P}_{\Sigma^3 W} X & \longrightarrow & \mathbf{P}_{\Sigma^2 W} X \\
\downarrow & & \downarrow & & \downarrow \\
* & \longrightarrow & \mathbf{P}_{\Sigma W} X & \longrightarrow & \mathbf{P}_{\Sigma W} X
\end{array}$$

Since by Proposition F.4 just proven the fibre $\mathbb{F}_1$ and $\mathbb{F}_2$ are GEMs then by definition $\mathbb{F}_3$ is a (twisted) polyGEM. This argument carries over by induction to show:

F.5 PROPOSITION: *The homotopy fibre $\mathbb{F}_N$ of the map σ_N is a polyGEM.*

But we now use (A.5) to show

F.6 PROPOSITION: *The homotopy fibre $\mathbb{F}_N$ of the map σ_N is an oriented polyGEM.*

Proof: We show that the inductive construction of the polyGEM uses principal fibrations. This follows from the fact that the fibres $\mathbb{G}_i$ in the tower

(F.6.1)
$$\begin{array}{ccccccccc}
\mathbb{G}_N & & \mathbb{G}_{N-1} & & & & \mathbb{G}_2 & & \\
\downarrow & & \downarrow & & & & \downarrow & & \\
\mathbf{P}_{\Sigma^N W} X & \longrightarrow & \mathbf{P}_{\Sigma^{N-1} W} X & \longrightarrow & \cdots & \longrightarrow & \mathbf{P}_{\Sigma^2 W} X & \longrightarrow & \mathbf{P}_{\Sigma W} X
\end{array}$$

(where $\mathbb{G}_i = \mathrm{Fib}(\mathbf{P}_{\Sigma^i X} X \longrightarrow \mathbf{P}_{\Sigma^{i-1} X} X)$) are all connected GEM spaces that have homotopically discrete spaces of pointed self-maps as was shown in (F.4.1) above, and thus of course discrete spaces of pointed self-homotopy equivalences. Their connectivity follows e.g. from (1.H.2) since each of the horizontal maps is a nullification with respect to a connected space. Since the base spaces of all the fibrations in the tower are 1-connected the classifying map can be lifted to the classifying space $\overline{W}\mathbb{G}_i = B\mathbb{G}_i$, which exactly means that these fibrations are principal. Stated otherwise for each $\mathbb{G}_i$ denoted here generically by $\mathbb{G}$ the fibration from which the universal fibration with a connected fibre $\mathbb{G}$ arise [D-Z-1], see (1.F.1):

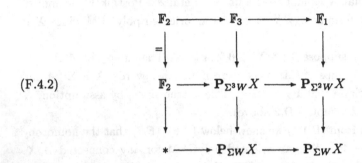

$$\mathrm{aut}^\bullet\mathbb{G} \longrightarrow \mathrm{aut}\mathbb{G} \longrightarrow \mathbb{G}$$

is a covering space fibration, since the fibre is discrete. But the base is 1-connected, so the total space is a disjoint union of copies of the base space $\mathbb{G}$ and as a group it is a twisted product. thus we have a fibration:

$$\mathrm{B}\mathbb{G} \longrightarrow \mathrm{Baut}\mathbb{G} \longrightarrow \mathrm{K}(\pi_0\mathrm{aut}^\bullet\mathbb{G}, 1)$$

Recall [May-1] that fibration with fibre $\mathbb{G}$ are classified by homotopy classes of maps to $\mathrm{Baut}\mathbb{G}$. Since the spaces in the tower above are simply connected the classifying maps to $\mathrm{Baut}\mathbb{G}$ can be lifted to $\mathrm{B}\mathbb{G}$, rendering the fibrations in the tower above principal as needed. ∎

The following is an immediate

F.7 COROLLARY: *For any connected space X the nullification $\mathbf{P}_{\Sigma^N X}\Sigma X$ is an oriented polyGEM.* ∎

Neisendorfer's Theorem on connected covers of finite complexes

Using the nullification functor and Miller's theorem, Neisendorfer has shown [N] that one can recover the p-completion of, say, a 2-connected finite complex from any high connected cover of this complex. Since the nullification of a point is always $\simeq$ to a point, we get that, if any high covering is contractible, then the original space must have a trivial p-completion. In particular, if one takes a sphere we detect in this way the non-triviality of p-torsion elements in infinitely high dimensions. Another result in the same spirit was discovered by McGibbon [McG] says that under mild conditions a finite space X can be recovered up to homotopy from the homotopy fibre of its canonical map to its stabilization QX.

Here is one formulation of this remarkable theorem [N].

F.8 THEOREM: *If X is a simply connected finite complex with $\pi_2(X)$ finite and $W = K(\mathbb{Z}/p\mathbb{Z}, 1)$, then for any integer $n \geq 0$ the nullification $\mathbf{P}_W(X_p < n >)$, of the p-completion of the n-th connected cover of X, has X_p itself as its p-completion.*

Proof: The proof is similar to that of (F.1). One interprets the Sullivan conjecture as saying that finite complexes are W-null and therefore do not change under nullification. On the other hand, the nullification of the fibre of the map from a high cover of X to X is killed by $\mathbf{P}_W$, assuming everything is p-complete, since these fibres are polyGEMs and so one can use (1.H.1) inductively on the finite construction of this polyGEM out of Eilenberg–Mac Lane spaces.

The following interesting variant of (F.8) is due to McGibbon [McG]:

F.9 THEOREM: *If X is as in (F.8) above then the map*

$$\mathbf{P}_W \mathrm{Fib}(X \longrightarrow \Omega^\infty \Sigma^\infty X) \longrightarrow \mathbf{P}_W X \simeq X$$

induces an equivalence on the p-completions [B-K], where $W = K(\mathbb{Z}/p\mathbb{Z}, 1)$ and p is any prime number.

We will not reproduce the proof here but remark that the theorem follows from another result in [McG] that shows that the nullification $\mathbf{P}_W$ kills up to p-completion any connected infinite loop space whose fundamental group is a torsion group. ∎

6. HOMOLOGICAL LOCALIZATION NEARLY PRESERVES FIBRATIONS

A. Introduction, main result

We now turn to homological localization $\mathbf{L}_{E\mathbb{Z}/p}$ for à generalized (non-connected) homology theory and, in particular, to mod-p complex K-theory and higher Morava K-theories. This section is largely taken from [DF-S]. Homological localization as defined by Bousfield [B-1] is a special case of homotopy localization with respect to a map $f: V \to W$.

Our concern in this chapter is to prove that in certain not-too-restrictive cases, Bousfield homological localization 'nearly' preserves fibration sequences. Notice the well known counterexample: Consider the fibre sequence

$$K(\mathbb{Z},2) \to * \to K(\mathbb{Z},3).$$

Since the base space in known to be K-acyclic [An-H] and the fibre $\mathbb{C}P^\infty = BS^1$, being a retract of BU, is K-local, the fibration is not preserved as such under K-localization. It turns out that in a certain precise sense this fibration plus a few others (see (C.2) below) exhaust the possible 'non-exactness' of K-theory localization for loop fibrations. This is because the general fibration theorem (5.D.6) above specializes for homological localization to yield a rather small 'error term' when localizing double loop of a fibration sequence. It is very probable that the same theorem holds even for single loop spaces. This near preservation of fibration is of course very useful by itself. It will be used in the next chapter to show that under not too restrictive assumptions there is an accessible description of the Bousfield localization with respect to mod-p K theory in terms of a certain telescope. It will of course render the mod-p homotopy groups of that localization more accessible.

A.1 THEOREM: *Let $\mathbf{L}_K$ be the homological localization with respect to complex mod-p K-theory. Let $F \to E \to B$ be a fibration over a 2-connected pointed space B. The homological localization $\mathbf{L}_K$ nearly preserves the fibration $\Omega^2 F \to \Omega^2 E \to \Omega^2 B$ up to an error term $J = K(H,2) \times K(G,1) \times K(S,0)$, where H is torsion free and G, S are abelian.*

Let $F \to E \to B$ be a fibration and $\mathbf{L}_{K(n)}$ the homological localization functor with respect to Morava K-theory $K(n)$, where $n \geq 0$. Up to an error term with at most three homotopy groups in dimensions $n-1, n, n+1$ the functor $\mathbf{L}_{K(n)}$ preserves the double loop of the given fibration. More generally, for any p-torsion homology theory $E\mathbb{Z}/p\mathbb{Z}$, there exists an integer $d(E)$, $1 \leq d \leq \infty$, such that $\mathbf{L}_{E\mathbb{Z}/p}$ preserves *any double-loop fibration up to an error term with possibly non-trivial homotopy groups only in dimension $d-1, d, d+1$.*

A.1.1 REMARK & NOTATION: The proof is given in section C below. Standard examples show that these three 'homotopy groups of the error term' do in fact arise non-trivially in homological localization. See (C.2) below.

We may assume that all the spaces involved are $H\mathbb{Z}/p$-complete with respect to usual homology. Since we work with a given prime p and HR-localization preserves any fibration which is a double-loop fibration by the fibre lemma of [B-K], since it is principal, therefore one may take the p-completion before applying $\mathbf{L}_E$.

Condition (5.D.6) (3) leads us to consider E_*-acyclic spaces, i.e. spaces X with $\mathbf{L}_E X \simeq *$. It will be convenient to denote by $\mathbf{P}_E$ the 'nullification or localization functor with respect to E-acyclic spaces', namely $\mathbf{P}_E = \mathbf{P}_{C(g)}$ where $C(g)$ is the mapping cone of g appropriate to E_*. In other words $\mathbf{P}_E = \mathbf{P}_{\mathrm{Acy}(E)}$, where $\mathrm{Acy}(E)$ is the wedge of all pointed E_*-acyclic spaces with cardinality not bigger than $\tilde{E}_* S^0$. For example, if E is integral homology, then $\mathbf{P}_E$ is the plus construction of Quillen. Notice that $\mathbf{P}_E X \simeq * \Leftrightarrow \mathbf{L}_E X \simeq *$ and this distinguishes $\mathbf{L}_E$ from a general $\mathbf{L}_f$.

B. Localization of polyGEMs

The main tool we need for homological localization is Theorem 5.D.6 above. In the present case we will need to analyze conditions (1), (2), (3) of that Theorem and, in particular, to gain some knowledge of the localization of polyGEMs. In general one would hope for a generalization of (4.B.3) and (4.B.4) for polyGEMs, namely that when the functor $\mathbf{L}_f$ is applied to a polyGEM the result is still a polyGEM. This however is not known. Luckily we can do with a statement concerning the loop space of a polyGEM.

B.1 LEMMA: *Let W be a space such that for some integer n and for any abelian group G one has $\mathbf{P}_W K(G,n) \simeq *$. If $Y \in \mathcal{S}$ is an oriented polyGEM we have vanishing homotopy groups as follows: $\pi_i \mathbf{P}_W \Omega Y \simeq 0$, for all $i \geq n$.*

Remark: It is likely that also $\pi_i \mathbf{P}_W Y \simeq 0$ for $i \geq n$. But in our proof we need to loop down once.

B.2 COROLLARY: *Let E_* be any homology theory for which there exists an integer $n \geq 1$ such that for any abelian group G one has $\mathbf{P}_E K(G,n) \simeq *$ (i.e. $\tilde{E}_* K(G,n) = 0$). Let T be a polyGEM with ΩT being E_*-local. Then $\pi_i \Omega T \simeq 0$ for all $i \geq n$.*

Proof of B.2: By (A.1.1), for E_* we have a space W that satisfies the conditions of (B.1). But then $\mathbf{P}_W \Omega T \simeq \Omega T$ since ΩT is E_*-local, thus the conclusion follows from (B.1.)

Proof of B.1: First assume that Y is a GEM. Then ΩY and, by (4.B.4) above, $\mathbf{P}_W \Omega Y$ are also GEM spaces. Since $\mathbf{P}_W \Omega Y$ is W-null GEM, so is every retract of it. Now $K(\pi_i \mathbf{P}_W \Omega Y, i)$ for every i is a retract of $\mathbf{P}_W \Omega Y$, thus this Eilenberg–Mac Lane space is W-null. By assumption on W we get $\pi_i \mathbf{P}_W \Omega Y \simeq 0$, as needed.

We proceed by induction on the construction of the oriented polyGEM Y. We use the inductive filtration of the class of oriented polyGEMs given in (5.A.2). We assume by induction that the conclusion of (B.1) holds for all k-polyGEMs. Let $Y = W_{k+1}$ be a $(k+1)$-polyGEM with $W_{k+1} \to X \to X^1$ a fibration in which X, X^1 are k-polyGEMs. Consider the (non-fibration) sequence:

$$\mathbf{P}_W \Omega W_{k+1} \to \mathbf{P}_W \Omega X \to \mathbf{P}_W \Omega X^1.$$

By Theorem 5.A.5 above this is a fibration up to a $K(G,n)$-null GEM, say Δ. In other words it induces a fibration:

$$\Delta \to \mathbf{P}_W \Omega W_{k+1} \to \text{Fibre } (\mathbf{P}_W \Omega X \to \mathbf{P}_W \Omega X^1).$$

Now Δ is $K(G,n)$-null GEM and therefore, by the above argument, its homotopy groups vanish above dimension $n-1$. By the induction assumption the same holds for the base space of this last fibration sequence. Therefore, it is also true for the total space, as needed.

B.3 COROLLARY: *If $\mathbf{L}_f K(G,n) \simeq *$ for some f and all G, then $\pi_i \mathbf{L}_f \Omega^3 Y \simeq 0$ for $i \geq n$ for any oriented polyGEM Y.*

Remark: Again this is likely to be true without the looping.

Proof: By (5.D.6)(1) $\mathbf{L}_f \Omega^2 Y$ is again a polyGEM. However, since $\mathbf{L}_f K(G,n) \simeq *$ we have $\mathbf{L}_f = \mathbf{P}_{K(G,n)} \mathbf{L}_f$, or $\mathbf{L}_f \Omega^2 Y$ is a $K(G,n)$-null double loop space, and the result follows by (5.A.5) above.

In the following lemma we show how (1), (2) and (3) of (5.D.6) combine to render $\Omega\Delta$ a small oriented polyGEM. The main idea is that, if a space X is both E_*-local and E_*-acyclic, then it must be contractible. Since in (5.D.6) there is a 'shift', Δ can still have several (at most three) non-zero homotopy groups.

B.4 LEMMA: *For any p torsion homology theory $(E\mathbb{Z}/p)_*$ there exists a typical number $1 \leq d \leq \infty$ such that if Δ is any 1-connected space satisfying:*

(1) *$\Omega\Delta$ is an oriented polyGEM,*
(2) *Δ is $E\mathbb{Z}/p$-acyclic,*
(3) *$\Omega^2\Delta$ is $E\mathbb{Z}/p$-local,*

then $\pi_i \Delta \simeq 0$, except possibly for $i \in \{d+1, d+2, d+3\}$.

Further, the groups $\pi_i \Delta$ satisfy:

(4) *$K(\pi_{d+1}\Delta, d+1)$ is $E\mathbb{Z}/p$-acyclic,*

(5) $K(\pi_{d+3}\Delta, d+1)$ *is* $\mathbf{P}_{E\mathbb{Z}/p}$*-local.*

In [B-6, 6.1–6.4] the homological localization of $K(G,n)$s with respect to any homology theory with $\mathbb{Z}/p\mathbb{Z}$-coefficients $E\mathbb{Z}/p$ is given. We say that an abelian group G is $H\mathbb{Z}/p$-local if $K(G,n)$ is $H\mathbb{Z}/p$-local for some (and thus for all [B-K]) integers $n \geq 1$. We need the following result of Bousfield:

B.5 PROPOSITION [B-6,6]: *Let G be any abelian $H\mathbb{Z}/p$-local group. Then there exist $1 \leq m \leq \infty$ such that:*

$$\mathbf{L}_{E\mathbb{Z}/p}K(G,i) \simeq K(G,i) \quad \text{for } 1 \leq i \leq m,$$
$$\mathbf{L}_{E\mathbb{Z}/p}K(G,i) \simeq * \quad \text{for } m+2 \leq i < \infty.$$

B.5.1 REMARK: In the extreme case $m = \infty$ the space $K(G,i)$ is $E\mathbb{Z}/p$-local for all $0 \leq i < \infty$. It is clear from (3.A.1) above that if $\mathbf{L}_{E\mathbb{Z}/p}K(G,i) \simeq *$ for some $i = d$, then the same holds for all $i \geq d$. Thus every homology theory E_* has a transitional dimension (as explained in [B-6, 8.1]); from this dimension and higher all Eilenberg–Mac Lane spaces are E_*-acyclic.

B.5.2 EXAMPLE: For complex or real K-theory we have $m = 1$.

B.6 PROOF OF B.4: We use the framework of Proposition B.5 above. The proof is then divided into three parts. We first deal with the easy cases $m = \infty$ and $m = 1$. Then we proceed to $1 < m < \infty$.

($m = \infty$). If for our homology theory $E\mathbb{Z}/p$ one has $m = \infty$, then we are in (3.5), (6.2) of [B-6] and the $E\mathbb{Z}/p$-localization for any space is the same as $H\mathbb{Z}/p$-localization, since being $E\mathbb{Z}/p$-local is equivalent to being $H\mathbb{Z}/p$-local. But since we assume (3), that $\Omega^2\Delta$ is $H\mathbb{Z}/p$-local, we get from standard properties of $H\mathbb{Z}/p$-localization that Δ itself is $H\mathbb{Z}/p$ -local and thus Δ is $E\mathbb{Z}/p$-local. But by (2), Δ is also $E\mathbb{Z}/p$-acyclic. Thus $\Delta \simeq *$. Therefore in that case the theorem is proved.

($m = 1$) This means that $K(\mathbb{Z}/p\mathbb{Z}, 1)$ is also $E\mathbb{Z}/p$-acyclic and therefore [B-6, 4.1] $E\mathbb{Z}/p$ is the trivial homology theory $E\mathbb{Z}/p \simeq *$, hence, $\Omega^2\Delta$ being $E\mathbb{Z}/p$-local, this also means $\Omega^2\Delta \simeq *$. Since Δ is 1-connected we get Δ has at most one non-trivial homotopy in dimension 1, so the conclusion holds.

($1 < m < \infty$). We now get to the non-trivial case where the integer m in B.5 is neither 1 nor ∞. In this case the argument is more involved and occupies the rest of the present section B: Denote this integer specific to E by $d = d(E) = m$. We now use Corollary B.2 above, to show that higher homotopy groups of Δ vanish. We can apply (B.2) with $T = \Omega\Delta$ for the theory $E\mathbb{Z}/p$ and $n = d+2$, since we have $\mathbf{L}_{E\mathbb{Z}/p}K(G, d+2) \simeq * \simeq P_{E\mathbb{Z}/p}K(G, d+2)$. Since $\Omega T = \Omega^2\Delta$ is $E\mathbb{Z}/p$-local and $T = \Omega\Delta$ is a polyGEM we conclude that $\pi_i \Omega T \simeq \pi_i \Omega^2 \Delta \simeq 0$ for $i \geq d+2$. Therefore

(B.6.1) $$\pi_i \Delta \simeq 0 \quad \text{for all } i \geq d+4.$$

On the other hand, we have chosen d such that any $H\mathbb{Z}/p$-local $K(G,i)$ for $i \leq d$ is $E\mathbb{Z}/p$-local [B-6, 6.3, 6.4]. Therefore the d-Postnikov section of Δ, namely $P_d\Delta$, is $\mathbf{P}_{E\mathbb{Z}/p}$-local since it is a finitely-repeated extension of $\mathbf{P}_{E\mathbb{Z}/p}$-local spaces by $\mathbf{P}_{E\mathbb{Z}/p}$-local Eilenberg–Mac Lane spaces (1.A.8) (e.6).

Therefore, in the fibration

$$\tilde{\Delta} \to \Delta \to \mathbf{P}_d\Delta.$$

The base is $\mathbf{P}_{E\mathbb{Z}/p}$-local and the fibre, being d-connected, has homotopy groups in at most three dimensions $d+1, d+2$ and $d+3$, as we saw in (B.6.1) above. Now using (3.D.3)(2) to localize the fibration we apply $\mathbf{P}_{E\mathbb{Z}/p}$ to get

(B.6.2) $$\mathbf{P}_{E\mathbb{Z}/p}(\tilde{\Delta}) \to \mathbf{P}_{E\mathbb{Z}/p}\Delta \to \mathbf{P}_{E\mathbb{Z}/p}(\mathbf{P}_d\Delta) = \mathbf{P}_d\Delta;$$

the base in the original sequence being $\mathbf{P}_{E_*}$-local, this is still a fibration. We will now show that all the three spaces in the fibration (B.6.2) are contractible: First, since by our assumption $\mathbf{L}_{E\mathbb{Z}/p}\Delta \simeq *$, we get, using Remark A.1.1 above, $\mathbf{P}_{E\mathbb{Z}/p}\Delta \simeq *$. We claim that $\mathbf{P}_{E\mathbb{Z}/p}\tilde{\Delta} \simeq *$ and therefore also $P_d\Delta \simeq *$. Second, notice that by the definition of d we have $\mathbf{P}_{E\mathbb{Z}/p}K(G, d+i) \simeq *$ for all $i \geq 2$. Therefore, taking the $(d+1)$-connected cover fibration over $\tilde{\Delta}$ and using (1.H.1) we get $\mathbf{P}_{E\mathbb{Z}/p}\tilde{\Delta} = \mathbf{P}_{E\mathbb{Z}/p}K(G, d+1)$ with $G = \pi_{d+1}\tilde{\Delta} = \pi_{d+1}\Delta$.

Now by (4.B.4) above $\mathbf{P}_{E\mathbb{Z}/p}K(G, d+1)$ is a d-connected GEM. Now, from the above $\mathbf{P}_{E\mathbb{Z}/p}$-localized fibration (B.6.2) and from $\mathbf{P}_{E\mathbb{Z}/p}\Delta \simeq *$, we get that this d-connected GEM is the loop space over a d-Postnikov section. This is possible only if both $P_d\Delta$ and its loop space $\mathbf{P}_{E\mathbb{Z}/p}\tilde{\Delta} \simeq \mathbf{P}_{E\mathbb{Z}/p}K(G, d+1)$ are contractible. In sum all the spaces in (B.6.2) are contractible.

In other words, Δ has no homotopy groups below dimension $d+1$ and, in conjunction with the first part of the proof, has non-vanishing homotopy groups at most in the three dimensions $d+1, d+2, d+3$, as claimed.

Now we turn to the proof of (4) and (5) in (B.4). To prove (4) consider the fibration sequence

$$\Delta\langle d+2\rangle \to \Delta \to P_{d+1}\Delta$$

where P_{d+1} denotes the $(d+1)$-Postnikov section of Δ. By our assumption on d any $K(G, d+i)$ is $E\mathbb{Z}/p\mathbb{Z}$ acyclic for $i \geq 2$. Since we know that the $(n+1)$-connected cover $\Delta\langle d+2\rangle$ has homotopy groups only in dimensions $d+2$ and $d+3$, it must be $E\mathbb{Z}/p\mathbb{Z}$ acyclic. Now in the above fibration both total space and fibre are $E\mathbb{Z}/p\mathbb{Z}$-acyclic, so the base is $E\mathbb{Z}/p$-acyclic too ((1.H.1) and (A.1.1) above). But the base space $P_{d+1}\Delta$ is $K(\pi_{d+1}\Delta, d+1)$, since we know that the lower homotopy groups of Δ vanish. This proves (4).

Finally, to prove (5), we use the fibration sequence

$$K(\pi_{d+3}\Delta, d+1) \to \Omega^2\Delta \to P_d\Omega^2\Delta.$$

Again by our assumption on d every $K(G, d-i)$ is $E\mathbb{Z}/p\mathbb{Z}$-local for $i \geq 0$. Therefore $P_d\Omega^2\Delta$ is the total space of a fibration with $\mathbf{P}_{E\mathbb{Z}/p}$-local base and fibre, hence it is $\mathbf{P}_{E\mathbb{Z}/p}$-local by (1.A.8)(e.6) above.

Since we assume (3), that $\Omega^2\Delta$ is $E\mathbb{Z}/p$-local, we can conclude from the above fibration sequence that the fibre $K(\pi_{d+3}\Delta, d+1)$ is $\mathbf{P}_{E\mathbb{Z}/p}$-local too, as needed.

This completes the proof of (B.4). ∎

C. Localization with respect to Morava K-theories

We now prove Theorem A.1. Let K_* be mod-p K-theory also denoted here by $K\mathbb{Z}/p_*$. As far as localization goes, K_* is the first in the series of Morava K-theories $K(n)$. We consider these theories here at an odd prime in line with [R-W], which considers the value of the theories $K(n)$ on Eilenberg–Mac Lane spaces for odd primes.

Let $\mathbf{L}_K$ be Bousfield's homological localization at mod-p K-theory $K\mathbb{Z}/p_*$. The results above specialize to the effect of $\mathbf{L}_K$ on a double loop of a fibration sequence $\Omega^2F \to \Omega^2E \to \Omega^2B$ for a 2-connected space B. In that case we consider the 'error term', namely the homotopy fibre J of the map:

$$\mathbf{L}_K\Omega^2 f \to \text{fibre}(\mathbf{L}_K\Omega^2E \to \mathbf{L}_K\Omega^2B).$$

Since $\mathbf{L}_g\Omega^2Y \simeq \Omega^2\mathbf{L}_{\Sigma^2 g}Y$ we first consider $\mathbf{L}_{\Sigma^2 g}$. We claim that the 'error term' fibre J is of the form $J = \Omega^2\Delta$ where Δ satisfies conditions (1), (2), (3) of B.4 above. First notice that since every $K\mathbb{Z}/p$-local space is also $H\mathbb{Z}/p$-local, all our spaces here are $H\mathbb{Z}/p\mathbb{Z}$-local. Second, J is clearly a double loop space since, for any map $f\colon V \to W$ and any fibration $F \to E \to B$ over a 2-connected B, the canonical comparison map $\mathbf{L}_f\Omega^2F \to \text{fibre}(\mathbf{L}_f\Omega^2E \to \mathbf{L}_f\Omega^2B)$ is the double loop of

$$\mathbf{L}_{\Sigma^2 f}F \to \text{fibre}(\mathbf{L}_{\Sigma^2 f}E \to \mathbf{L}_{\Sigma^2 f}B);$$

so the fibre of that canonical comparison map is the double loop of the fibre of that latter map (3.A.1).

It now remains to recall that by [An-H] the space $K(G,n)$ is $K\mathbb{Z}/p\mathbb{Z}$-acyclic for all $n \geq 3$ and all G. On the other hand, by [B-6] $K(G,1)$ is $K\mathbb{Z}/p\mathbb{Z}$-local if it is $H\mathbb{Z}/p$-local and G is abelian. Therefore for mod-p K-theory one has $d(K\mathbb{Z}/p) = 1$, where d is as in (B.4). Therefore, besides π_0 the possible homotopy groups of the error term J above are in dimensions one and two. Thus J is a disjoint union of 2-stage Postnikov sections $J = \amalg J_0$ sitting in a fibration sequence:

$$K(\pi_2, 2) \to J_0 \to K(\pi_1, 1).$$

Now by (B.4) (5) we know that $K(\pi_2, 2)$ is $K\mathbb{Z}/p$-local. Therefore π_2 is $\mathbb{Z}/p\mathbb{Z}$-complete without $\mathbb{Z}/p\mathbb{Z}$-torsion, i.e. a free module over $\mathbb{Z}_p^\wedge$. Since J_0 is a double loop space, this fibration splits and $J_0 \simeq K(\pi_2, 2) \times K(\pi_1, 1)$; see Lemma C.1 below.

This concludes the proof for the case of K-theory.

We now turn to the higher Morava K-theories at a given odd prime p. By the calculations in [R-W, 12.1], for odd primes,

$$\tilde{K}(n)K(G, n+2) = 0 \text{ and } \tilde{K}(n)K(\mathbb{Z}/p\mathbb{Z}, n+1) = 0.$$

We can apply (B.4) above to conclude: If Y is the homotopy fibre of

$$\mathbf{L}_{K(n)}\Omega^2 F \to \text{fibre}(\mathbf{L}_{K(n)}\Omega^2 E \to \mathbf{L}_{K(n)}\Omega^2 B),$$

it fits into a fibration:

$$K(F, n+1) \times K(G, n) \to Y \to K(S, n-1)$$

where F is a torsion free group.

Again, Y being a double loop space and F torsion free, we can conclude that the possible K-invariant connecting the two higher groups must vanish, since there are no unstable elements in this range.

This concludes the proof of (B.4) for Morava K-theories at odd primes. ∎

C.1 LEMMA: *Let $Y \simeq \Omega^2 Y^1$ be a K-local space with ΩY^1 a polyGEM. Then Y_0, the null component of Y, fits into a fibration $K(F,2) \to Y_0 \to K(G,1)$ where F is a torsion free group.*

Proof: We use (B.2) below for $T = \Omega Y^1$. Since by [An-H] reduced K-theory vanishes on $K(\mathbf{Z},3)$, we get $\mathbf{L}_K K(\mathbf{Z},3) \cong *$. Applying (B.4) above we get that Y_0, our K-local polyGEM, fits into a fibration:

$$K(G',2) \to Y_0 \to K(G,1),$$

since it only has two non-vanishing homotopy groups. We claim that G' is a torsion free group. To see this we observe that both base and total space Y are $K(\mathrm{Tor},2)$ null with respect to any abelian torsion group Tor: For the base observe directly that $\mathrm{map}_*(K(\mathrm{Tor},2), K(G,1)) \simeq *$. And the total space Y is, by assumption, K-local, so it is null with respect to any K-acyclic space, such as $K(\mathrm{Tor},2)$ [An-H]. Therefore the fibre $K(G',2)$ must also be K-local and, in particular, $K(\mathrm{Tor},2)$-null, so it can admit no maps from $K(\mathrm{Tor},2)$. Hence Tor $G' \cong 0$ and G' is torsion free.

C.2 Standard examples: We now give a short list of examples that show how the three possibly non-trivial homotopy groups in the error term in (A.1) and (B.4) actually arise in standard fibration sequences. We consider the effect of $\mathbf{L}_K$ on fibrations:

1. In the fibration

$$\mathrm{SU}\mathbf{Z}/p \to * \to \mathrm{BSU}\mathbf{Z}/p$$

the error term is $K(\mathbf{Z}/p\mathbf{Z},0)$. This is true since the K-localization of the infinite loop space $\mathrm{BSU}\mathbf{Z}/p$ is homotopy equivalent to $\mathrm{BU}\mathbf{Z}/p$ using, say, [B-3] while $\mathrm{SU}\mathbf{Z}/p$ is K-local.

2. In the fibration

$$K(G,1) \to * \to K(G,2)$$

for any abelian G, the error term is $K(\mathrm{Tor}\ G,1)$.

3. In the fibration

$$K(\mathbf{Z},2) \to * \to K(\mathbf{Z},3)$$

the error term is $K(\mathbf{Z},2)$.

Thus the three possible dimensions actually arise. Similar examples can be given for the higher Morava K-theories.

7. CLASSIFICATION OF NULLITY AND CELLULAR TYPES OF FINITE p-TORSION SUSPENSION SPACES

Introduction

Let $\mathcal{F}_*^p$ be the full subcategory of $\mathcal{S}_*$ consisting of finite p-torsion spaces, namely W is finite and $H_*(W, \mathbb{Z})$ is a finite p-torsion group for $W \in \mathcal{F}_*^p$.

We will consider nullity classes $\langle - \rangle$ and cellular classes $\mathcal{C}^{\cdot}(-)$. It turns out that after a single suspension both classes in $\mathcal{F}_*^p$ can be understood in terms of Morava K-theories and, in particular, Hopkins–Smith theory of *types*. These two classifications are closely related in view of (3.B.3) above. Recall that $\langle W \rangle = \langle V \rangle$ or V and W have the same nullity type if $V < W$ and $W < V$ or $\mathbf{P}_V W \simeq *$ and $\mathbf{P}_W V \simeq *$, alternatively, if for all $X \in \mathcal{S}_*$ one has the double implication.

$$\text{map}_*(V, X) \simeq * \iff \text{map}_*(W, X) \simeq *.$$

WARNING: Here the partial order $<$ is used in the opposite sense to that of [B-4]. The present notation is consistent with connectivity i.e. $X < Y$ implies $\text{conn}X \leq \text{conn}Y$. Also $S^n < S^{n+1} < S^{n+2} \cdots$ etc. It is also consistent, as we will see here, with the Hopkins-Smith type.

Also recall that $\mathcal{C}^{\cdot}(A) = \mathcal{C}^{\cdot}(B)$ or A and B have the same cellular type if $A \ll B$ and $B \ll A$ or if $CW_A B \simeq B$ and $CW_B A \simeq A$. Both $<$ and $\ll$ define partial orders on $\mathcal{S}_*$ and $\mathcal{F}_*^p$. Our aim is to determine these partially ordered sets of types up to isomorphism. This can be done in $\Sigma\mathcal{F}_*^p$, the suspension spaces in $\mathcal{F}_*^p$. It turns out that in $\Sigma\mathcal{F}_*^p$ the above partial order can be reduced to K-theoretical and homological properties.

A. Stable nullity classes and Hopkins–Smith types

Associated with nullity and cellular types are the corresponding stable notions. It turns out that these agree with each other and with the notion of Hopkins–Smith type [H-S]. With each $W \in \mathcal{S}_*$ we have a (generally strictly) increasing series

(*) $$W < \Sigma W < \Sigma^2 W < \cdots < \Sigma^k W < \cdots,$$

(*) $$W \ll \Sigma W \ll \Sigma^2 W \ll \cdots \ll \Sigma^k W \ll \cdots.$$

Therefore we can define 'stable' versions of $<$ and $\ll$: namely $A <_s B$ (or $A \ll_s B$) if for some k $A < \Sigma^k B$ (or $A \ll \Sigma^k B$). In view of (3.B.3) above the partial orders $<_s$ and $\ll_s$ are the same: $A \ll_s B$ if and only if $A <_s B$. We denote in line with Bousfield the stable equivalence classes by $\{A\}$, namely $\{A\} = \{B\}$ if $A <_s B$ and $B <_s A$.

First let us recall (5.B.6) above. In the present context it says that on suspension spaces the partial order ($<$) is inherently stable:

COROLLARY 5.B.6: *For any spaces X, Y and integer $k \geq 1$ one has: $\Sigma X < \Sigma Y$ iff $\Sigma^k X < \Sigma^k Y$.*

But the main observation about stable types is the following relation between stable classes $\{A\}$ and Hopkins–Smith types:

A.1 THEOREM: *For any prime p and spaces $W, V \in \mathcal{F}_*^p$ one has $\{A\} <_s \{B\}$ if and only if type $A \leq$ type B.*

Recall: For each space $X \in \mathcal{F}_*^p$ with $\tilde{H}_*(X, \mathbb{Z}) \neq 0$ there exists an integer $n \geq 0$ for which $\tilde{K}(m)_* X = 0$ for all $m < n$ and $\tilde{K}_*(m) X \neq 0$ for all $m \geq n$, where $\tilde{K}(m)_*$ is the mth reduced Morava K-theory at the prime p. This integer n is called the type of X at the prime p. For a clear survey see [H], [R]. Further, the type of X is related to classification by Hopkins and Smith of so-called thick sub-categories. A full subcategory $\emptyset \neq U \subset \mathcal{F}_*^p$ is called thick if it is closed under the third term in a cofibration sequence $A \hookrightarrow X \to X/A$ and under homotopy retracts. Their main theorem classifies all thick sub-categories:

A.2 THEOREM (Thick subcategory): *If $\mathcal{U}$ is a thick subcategory in F_*^p, then $\mathcal{U}$ is equal to the thick subcategory of all spaces of type $\geq n$ for some $0 \leq n \leq \infty$.*

Using $<_s$ we can define for each space $W \in F_*^p$ a thick subcategory as follows:

A.3 DEFINITION: *The thick subcategory associated with W, denoted by $\mathcal{U}(W)$, consists of all $V \in F_*^p$ for which $W <_s V$.*

Proof of A.1: We must show that $\mathcal{U}(W)$ is closed under retracts and third terms in cofibrations. Since for any retract V' of V one has $V < V'$, and thus $V <_s V'$, the first is clear. Using the Puppe sequence $A \hookrightarrow X \to X/A \to \Sigma A$ and (3.D.1) above, the second closure property follows immediately: $\mathbf{P}_W A \simeq *$ and $\mathbf{P}_W X \simeq *$ implies $\mathbf{P}_W(X/A) \simeq *$ etc. On the other hand, if $\tilde{K}(m)_* A \simeq 0$ and $\mathbf{P}_A \Sigma^k B \simeq *$, then clearly $\tilde{K}(m)_* \Sigma^k B \simeq 0$, thus B is also $\tilde{K}(m)_*$-acyclic. So we have type $A \leq$ type B if $A <_s B$.

To conclude, one can reformulate the above discussion as follows:

A.4 THEOREM: *The following classifications of $\mathcal{F}_*^p$ are one and the same as partially ordered sets of equivalent classes:*

(1) *stable classes of nullity types,*
(2) *stable classes of cellular types,*
(3) *thick sub-categories of $\mathcal{F}_*^p$.*

B. Unstable nullity types

Without stabilization as above there is a difference, 'at the bottom dimension', between nullity type and cellular type of spaces in $\mathcal{S}_*$ or in $\mathcal{F}_*^p$. The canonical example comes from (3.C.9) and (2.D.11), namely $K(\mathbb{Z}/p^2\mathbb{Z};n)$ and $K(\mathbb{Z}/p\mathbb{Z},n)$ have the same nullity type while they do not have the same cellular type. Rather we have $K(\mathbb{Z}/p^2\mathbb{Z},n) \ll K(\mathbb{Z}/p\mathbb{Z},n)$, but the opposite cellular inequality does not hold.(Compare [Bl-2, 3.1].)

B.1 EXAMPLE: Since we have a fibration sequence for each $n \geq 1$,

$$K(\mathbb{Z}/p^2\mathbb{Z},n) \xrightarrow{\times p} K(\mathbb{Z}/p^2\mathbb{Z},n) \to K(\mathbb{Z}/p,n+1) \times K(\mathbb{Z}/p,n),$$

it follows from (3.C.9) and (2.D.11) above that the product at the base of the fibration and each of its factors is a $K(\mathbb{Z}/p^2\mathbb{Z},n)$-cellular space. Thus $K(\mathbb{Z}/p^2\mathbb{Z},n) \ll K(\mathbb{Z}/p\mathbb{Z},n)$. Similarly, one has for Moore spaces $M^d(G)$ the cellular inequalities $M^d(\mathbb{Z}/p^2\mathbb{Z}) \ll M^d(\mathbb{Z}/p\mathbb{Z})$ but not in the opposite sense. Using the technique of Chapter 5 the classification of nullity type can be reduced to the stable classification for $\Sigma\mathcal{F}_*^p$, i.e. suspension spaces in $\mathcal{F}_*^p$.

NOTATION Let conn$X$ denote the connectivity of X.

B.2 THEOREM [B-4]: *For a given prime p let W, V be two spaces in $\mathcal{F}_*^p$. Then the following two conditions are equivalent:*

(i) $$\Sigma W < \Sigma V,$$

(ii) $$\Sigma W <_s \Sigma V \quad \text{and} \quad \text{conn}\Sigma W \leq \text{conn}\Sigma V.$$

Proof: The implication (i) ⇒ (ii) is the easier one. Assuming (i), then $\Sigma W <_s \Sigma V$ is immediate while conn$\Sigma W \leq$ connΣV follows from:

B.3 PROPOSITION: *For any $A, B \in \mathcal{S}_*$, if $A < B$ then* conn$A \leq$ conn$B$.

Proof: Assume, by negation, that conn$A >$ conn$B = n$. Then for dimensional reasons it follows that for every group G we have $\text{map}_*(A, K(G,n+1)) \simeq *$. But the canonical map $B \to K(\pi_{n+1}B, n+1)$ is not null if conn$B = n$ (where *conn* denotes the connectivity), so $\mathrm{map}_*(B, K(\pi_{n+1}B, n+1)) \not\simeq *$. This contradicts $\mathbf{P}_A B \simeq *$. The proof of the implication (i) $\Rightarrow$ (ii) is now complete.

To prove the opposite implication we follow the approach of Bousfield in identifying the nullity type of a finite space $\Sigma X \in \mathcal{S}_*^p$ with that of a certain infinite space $\Sigma^k X \vee K(\mathbb{Z}/p\mathbb{Z}, d)$ for any $k \geq 0$ and $d = \mathrm{conn}\Sigma X + 1$.

Assuming (B.4) below the proof is easily completed: We must show that for *some* integers k, l there is an inequality:

$$\Sigma^k W \vee K(\mathbb{Z}/p, n+2) \;<\; \Sigma^l V \vee K(\mathbb{Z}/p, m+2)$$

where $n = conn\Sigma W, m = conn\Sigma V$. But this is clear since by assumption $n \leq m$ so $K(\mathbb{Z}/p, n+2) \;<\; K(\mathbb{Z}/p, m+2)$ and also by choosing l large enough in relation to k one gets $\Sigma^k W \;<\; \Sigma^l V$. This completes the proof of (B.2). ∎

In the proof of (B.2) we have used the following basic result is due to Bousfield. We give here a somewhat different proof based on Chapter 4 and the useful (B.4,B.5) below. For simplicity we consider here only finite p-torsion spaces; a similar statement holds for all p-torsion spaces.

B.4 THEOREM: *Let $X \in \mathcal{F}_*^p$ be a finite, p-torsion space and let $n = \mathrm{conn}X$. Then for each $k \geq 1$ the spaces ΣX and $Y = \Sigma^k X \vee K(\mathbb{Z}/p, n+2)$ have the same nullity type.*

Proof of (B.4): The proof runs over the next two pages using several propositions and lemmas of independent interest. We must show $\mathbf{P}_{\Sigma X} Y \simeq *$ and $\mathbf{P}_Y \Sigma X \simeq *$. Consider the first and easier part. Since $k \geq 1$ we have $\mathbf{P}_{\Sigma X} Y \simeq \mathbf{P}_{\Sigma X} K(\mathbb{Z}/p\mathbb{Z}, n+2)$ by (*) in (A) above.

By Proposition B.5 below one has

$$\mathbf{P}_{\Sigma X} K(\mathbb{Z}/p\mathbb{Z}, n+2) \simeq \mathbf{P}_{\mathbb{Z}\Sigma X} K(\mathbb{Z}/p\mathbb{Z}, n+2).$$

If $G = H_{n+1}(X, \mathbb{Z})$, then $K(G, n+2)$ is, up to homotopy, the bottom factor in the GEM product $\mathbb{Z}\Sigma X$. Thus $\mathbf{P}_{\Sigma X} Y \simeq *$ will follow from $\mathbf{P}_{K(G,n+2)} K(\mathbb{Z}/p, \mathbb{Z}, n+2) \simeq *$. Since by assumption G is a non-trivial p-torsion group, the last equivalence follows from our example (3.C.9) above, proving $\mathbf{P}_{\Sigma X} Y \simeq *$.

Now consider $\mathbf{P}_Y \Sigma X$. By (5.F.7), $\mathbf{P}_{\Sigma^k X} \Sigma X$ is an oriented polyGEM with p-torsion homotopy groups. In fact by (C.7) below it is a finite Postnikov stage with a finite number of finite p-torsion homotopy groups. This polyGEM is $(n+1)$-connected and thus the desired $\mathbf{P}_Y \Sigma X \simeq *$ follows from the following lemma (B.4.1).

B.4.1 LEMMA: *If $W \in \mathcal{F}^p$ is an m-connected finite p-torsion polyGEM then* $\mathbf{P}_{K(\mathbb{Z}/p\mathbb{Z},m+1)}W \simeq *$.

Proof: This is true if W is an m-connected p-torsion GEM. Now use (1.H.1) to proceed by induction on the finite construction of the polyGEM. ∎

B.5 PROPOSITION: *For any GEM space X and any $W \in \mathcal{S}_*$, there is a natural homotopy equivalence*

$$\mathbf{P}_{\mathbb{Z}W}X \xrightarrow{\simeq} \mathbf{P}_W X.$$

We can reformulate this proposition as follows:

B.5.1 PROPOSITION: *For any GEM space X and any $W \in \mathcal{S}_*$, we have the double implication:*

$$\mathrm{map}_*(W, X) \simeq * \Leftrightarrow \mathrm{map}_*(\mathbb{Z}W, X) \simeq *.$$

These two propositions are one and the same, since we know a priori that if X is a GEM then the spaces $\mathbf{P}_{\mathbb{Z}W}X$ and $\mathbf{P}_W X$ are both GEM spaces, and we use the following lemma in conjunction with $W < \mathbb{Z}W$ (since, in fact, $W \ll \mathbb{Z}W$); see Chapter 4. In fact all one needs to show in order to get the homotopy equivalence is that $\mathbf{P}_{\mathbb{Z}W}X$ is W-null if X is a GEM.

B.6 LEMMA: *If $W < V$, then we have natural equivalences of functors*

$$\mathbf{P}_W \simeq \mathbf{P}_W\mathbf{P}_V \simeq \mathbf{P}_V\mathbf{P}_W.$$

Proof: Given the first equivalence, the second follows since $\mathbf{P}_W X$ is W-null and thus V-null for all X. To get the first equivalence we start with the map $\mathbf{P}_W X \to \mathbf{P}_W\mathbf{P}_V X$, which is $\mathbf{P}_W$ applied to the coaugmentation. To get a homotopy inverse $\mathbf{P}_W\mathbf{P}_V X \to \mathbf{P}_W X$ we start with $X \to \mathbf{P}_W X$. Since $W \ll V$ this is a map to a V-null space, and it factors uniquely up to homotopy through $X \to \mathbf{P}_V X \to \mathbf{P}_W X$. But now the map on the right is a map of $\mathbf{P}_V X$ into a W-null space, so it factors through $\mathbf{P}_V X \to \mathbf{P}_W\mathbf{P}_V X \to \mathbf{P}_W X$. Homotopy uniqueness now gives that the map $\mathbf{P}_W\mathbf{P}_V X \to \mathbf{P}_W X$ we have obtained is the desired homotopy inverse.

Proof of Proposition B.5.1: One implication follows immediately from $W \ll \mathbb{Z}W$ as in (4.A.2.1) above. To get the other direction we use the basic lemma (4.B.2) above: Any map $W \to$ GEM from W to a GEM space factors up to homotopy through $W \to \mathbb{Z}W \to$ GEM. Since we assume $\mathrm{map}_*(\mathbb{Z}W, X) \simeq *$, we can conclude $[W, X] \simeq *$. Therefore, to get $\mathrm{map}_*(W, X) \simeq *$ it is sufficient to show $[\Sigma^k W, X] \simeq *$ for every integer $k \geq 0$. Since X is a GEM, it is sufficient to show that $[\mathbb{Z}\Sigma^k W, X] \simeq$

$*$. Since we are given $\text{map}(\mathbb{Z}W, X) \simeq *$, we know that $[\Sigma^k \mathbb{Z}W, X] \simeq *$. Therefore our claim follows from

B.7 LEMMA: *For any $k \geq 0$ and $W \in \mathcal{S}_*$ we have $\Sigma^k \mathbb{Z}W \ll \mathbb{Z}\Sigma^k W$.*

Proof: In fact $\mathbb{Z}\Sigma^k W$ is obtained from $\mathbb{Z}W$ by taking k-times the classifying space functor $\mathbf{B}^k \mathbb{Z}W \simeq \mathbb{Z}\Sigma^k W$. The last equivalence follows directly from the basic property of $\mathbb{Z} \simeq SP^\infty$, the infinite symmetric product functor as investigated by Dold and Thom [D-Th]; namely if $A \to X \to X/A$ is any cofibration, then $\mathbb{Z}A \to \mathbb{Z}X \to \mathbb{Z}(X/A)$ is a fibration sequence. Therefore (letting $\mathbf{B}$ denote the classifying space functor) $\mathbb{Z}X \to * \to \mathbb{Z}\Sigma X$ is a fibration sequence; in other words $\mathbb{Z}\Sigma X = \mathbf{B}\mathbb{Z}X$. But we will show in Chapter 9 below (9.D.3) that $\mathbf{B}G \gg \Sigma G$ for any group object in $\mathcal{S}_*$ and thus, by induction, $\mathbf{B}^k G \gg \Sigma^k G$ for any abelian group object in $\mathcal{S}_*$. This completes the proof of the B.7 and thus of Theorem B.4.

∎

An immediate deduction is

B.8 COROLLARY: *The following two conditions are equivalent in $\mathcal{F}^p_*$:*

(i) *ΣX and ΣY have the same nullity,*

(ii) *type ΣX = type ΣY and* $\text{conn}\Sigma X = \text{conn}\Sigma Y$.

C. Unstable cellular types

The classification of cellular types is very closely related to that of nullity types. We saw above that, while the two Moore spaces $M^d(\mathbb{Z}/p)$ and $M^d(\mathbb{Z}/p^2\mathbb{Z})$ have the same nullity, they are not of the same cellular type so that the inequality $M^d(\mathbb{Z}/p\mathbb{Z}) \ll M^d(\mathbb{Z}/p\mathbb{Z}^2)$ is strict.

The classification is related to the p-torsion in the bottom dimension of ΣX [Bl-2].

C.1 DEFINITION: *Let G be a finite abelian p-torsion group. We denote by $t(G)$ the exponent of G i.e. the maximal integer ℓ for which $\mathbb{Z}/p^\ell\mathbb{Z}$ is a direct summand of G or the minimal ℓ with $p^\ell \cdot G = 0$.*

With this integer $t(G)$ we can formulate the classification of cellular types in ΣF^p_* as follows:

C.2 THEOREM: *Let $X \in \mathcal{F}^p_*$ and let $n = \text{conn}X$ and $\ell = t(H_{n+1}(X, \mathbb{Z}))$. Then for each $k \geq 0$ the spaces ΣX and $\Sigma^k X \vee K(\mathbb{Z}/p^\ell\mathbb{Z}, n+2)$ have the same cellular types.*

Proof: The proof is very similar to that of Theorem (B.4) above, (Compare [Bl, 3.1] except that we use the following version of Proposition (B.5).

C.3 PROPOSITION: *If X is a GEM and A is any space in $\mathcal{S}_*$, then there is a natural homotopy equivalence*

$$e : CW_{\mathbb{Z}A}X \overset{\simeq}{\to} CW_A X.$$

Proof: Since $A \ll \mathbb{Z}A$ by (4.A.2.1) above we conclude $CW_{\mathbb{Z}A}(CW_A X) \simeq CW_{\mathbb{Z}A}X$, hence the natural map. To get an inverse we use (C.4) below: Since by (4.B.3) $CW_A X$ is a GEM and also A-cellular, it is a $\mathbb{Z}A$-cellular space by Proposition C.4 below. Therefore the natural augmentation $CW_A X \to X$ factors up to homotopy through $CW_{\mathbb{Z}A}X \to X$, which is the universal map of $\mathbb{Z}A$-cellular spaces to X. Hence we get a map $f : CW_A X \to CW_{\mathbb{Z}A}X$. Universality and uniqueness up to homotopy now imply as usual that f is a homotopy inverse to e.

We have used

C.4 PROPOSITION: *If X is an A-cellular GEM, then X is also $\mathbb{Z}A$-cellular.*

Proof: This follows from Proposition (C.5) below about symmetric products. Since $A \ll X$ by assumption, we get $\mathbb{Z}A \ll \mathbb{Z}X$ by (C.5). But X is a GEM, so X is a retract, up to homotopy, of $\mathbb{Z}X$; thus $\mathbb{Z}X \ll X$ by (2.D.1 (5)) above. It follows that $\mathbb{Z}A \ll X$, as required.

C.5 PROPOSITION: *For any $A, X \in \mathcal{S}_*$, if $A \ll X$ then $\mathbb{Z}A \ll \mathbb{Z}X$.*

Proof: (Arguing simplicially.) We proceed by induction on the construction of X from A, namely assume X is given as a pointed homotopy colimit by $\int_I A_i$ and $A_i \gg A$, where we assume by induction that $\mathbb{Z}A_i \gg \mathbb{Z}A$ and prove $\mathbb{Z}X \gg \mathbb{Z}A$.

C.6 LEMMA: *If Y is a free I-diagram in $\mathcal{S}_*$, then*

$$\mathbb{Z}\int_I Y \simeq \operatorname{dirlim}_I \mathbb{Z}Y$$

where the direct limit is taken in the category of abelian groups.

Proof: This is immediate, since for a free diagram Y in $\mathcal{S}_*$ the homotopy colimit colim $\int_I Y$ is just $\operatorname{colim}_I Y$ and the free abelian group functor $\mathbb{Z}$ [B-K], being left adjoint, commutes with colimits (= direct limits).

To proceed with the proof we write $\mathbb{Z}X$ as $\mathbb{Z}\int_I A_i$.

Now by cellular induction we may assume that the category I is either discrete, so that $\operatorname{colim}_I A_i = \bigvee A_i$, or a coequalizer diagram, so that $I = \cdot \rightrightarrows \cdot$. By our lemma $\mathbb{Z}\int A_i$ is isomorphic to $\varinjlim \mathbb{Z}A_i$ in the category of abelian groups, where the direct limit is taken over the appropriate small category. To proceed with the proof of

(C.5) we examine each of the two small categories I. In both cases we assume by cellular induction that $A_i \gg A$ have the desired property, namely that $\mathbb{Z}A_i \gg \mathbb{Z}A$.

Let us start with the second, coequalizer case: We may assume that $(A_i) = (A_1 \overset{f_1}{\underset{f_2}{\rightrightarrows}} A_2)$ is free, i.e. f_1 and f_2 are cofibrations that agree only on the base point. In that case $\mathrm{colim}_i \mathbb{Z}A_i$ in the category of abelian groups is given by a short exact sequence of simplicial abelian groups

$$\mathbb{Z}A_1 \overset{f_1-f_2}{\hookrightarrow} \mathbb{Z}A_2 \to \mathbb{Z}X = \mathrm{colim}\mathbb{Z}A_i.$$

Thus we have a principal fibration sequence with the base $\mathbb{Z}X$ and, by induction, both fibre and total space are in $C^{\cdot}(A)$. It follows by (2.D.11) above that the base $\mathbb{Z}X \in C^{\cdot}(A)$, as needed.

Turning now to the first case of a wedge the direct limit in the category of abelian groups is given by $\mathbb{Z}X = \oplus\mathbb{Z}A_i$. Again we assume $\mathbb{Z}A_i \gg A$ and must show that $\oplus_i \mathbb{Z}A_i \gg A$. But $\oplus_{i\in I}\mathbb{Z}A_i = \mathrm{colim}_{(\alpha)}(\underset{i<\alpha}{\oplus} \mathbb{Z}A_i)$ where (α) represents a well ordering of I. Thus since $C^{\cdot}(A)$ is closed under linear direct limits which now are taken in $\mathcal{S}$ and finite products, we get the desired result.

Special cases: n-supported W, p-torsion W

The functor $\mathbf{P}_{\Sigma W}$ is best behaved when we assume a certain relation between its first non-trivial and higher homology groups. We would like to know when the error terms have at most one non-trivial homotopy group. It turns out that this can be guaranteed if we assume that $K(H_n, W, n)$ satisfies

$$K(H_n W, n) < K(H_{n+i} W, n)$$

where $H_n W$ is the first non-trivial integral homotopy and $i \geq 0$. Alternatively, we say that W is *n-supported* (compare [B-4, 7.1]) if $\mathbf{P}_{P_n \mathbb{Z}W}\mathbb{Z}W \simeq *$, where $n = \mathrm{conn} H_* W + 1$. In other words, W is n-supported if $P_n\mathbb{Z}W < \mathbb{Z}W$.

C.7 Proposition: *Let W be an n-supported space. Assume that in the fibration $F \to E \to X$ we have $\mathbf{P}_{\Sigma W} E \simeq \mathbf{P}_{\Sigma W} X \simeq *$. Then $\mathbf{P}_{\Sigma W} F$ has a single homotopy group in dimension $n = \mathrm{conn} H_* W + 1$.*

C.7.1 Example: If the n-homotopy group of W has the group of integers $\mathbb{Z}$ as a retract, then W is n-supported. The reason is, in that case, $K(H_n W, n) < K(\mathbb{Z}, n)$ and for any G we have $K(\mathbb{Z}, n) < K(G, n+i)$ for all $i \geq 0$.

C.7.2 Example: [B-4]: If $\tilde{H}_i(W, \mathbb{Z}) \simeq 0$ for $i < n$ and $\tilde{H}_*(W, \mathbb{Z})$ is J-torsion for some set of primes (resp. J-local) and $H^n(W, \mathbb{Z}/p) \cong \mathrm{Hom}(H_n(W, \mathbb{Z}), \mathbb{Z}/p)$ is not trivial for each $p \in J$ (resp. $H^n(W, \mathbb{Z}_{(J)}) \not\simeq 0$), then W is n-supported. This is true, since the existence of a non-trivial, and thus surjective, map $H_n(W, \mathbb{Z}) \to \mathbb{Z}/p$ guarantees that $K(H_n(W,\mathbb{Z}), n) < K(\mathbb{Z}/p, n)$, for each $p \in \mathcal{J}$. Since $H_{n+i}W$ are $\mathcal{J}$-torsion groups by assumption, we get

$$K(\mathbb{Z}/p, n) < K(\mathbb{Z}/p\mathbb{Z}, n+i) < K(H_{n+i}(W, \mathbb{Z}), n+i),$$

for all $i \geq 0$ as needed (resp. for $\mathcal{J}$-local).

C.7.3 Example: If $H_k(W, \mathbb{Z})$ are all p-torsion for a fixed prime p and $H_n(W, \mathbb{Z})$ is finitely generated, then W is n-supported. In this case $H_n(W, \mathbb{Z})$ contains a finite p-torsion group $\mathbb{Z}/p^\ell\mathbb{Z}$ as a direct summand. Therefore

$$K(H_n W, n) < K(\mathbb{Z}/p^\ell \mathbb{Z}) < K(\mathbb{Z}/p\mathbb{Z}, n) < K(H_i W, n+i)$$

for each $i \geq 0$, since $H_i W$ is a p-torsion group.

Proof of Proposition C.7: As we have seen in Chapter 4 above, in such a fibration $\mathbf{P}_W F \simeq *$, F is ΣW-null and F is a GEM. Therefore by (B.5), $\mathbf{P}_{\mathbb{Z}W} F \simeq *$. But we assume $P_n \mathbb{Z}W < \mathbb{Z}W$. Therefore $\mathbf{P}_{P_n\mathbb{Z}W} F \simeq *$. On the other hand F is ΣW-null, and since $\Sigma W < \mathbb{Z}\Sigma W = \overline{W}\mathbb{Z}W$ (where $\overline{W}$ is the classifying space functor) we get that F is $\overline{W}\mathbb{Z}W$-null; in particular it is $K(H_n W, n+1)$-null, the latter space being a retract of $\overline{W}\mathbb{Z}W$. But since $K(H_n W, n) < F$ we get $K(H_n W, n+1) < F\langle n\rangle$ (the n-connected cover of F). Now we obtain that $F\langle n\rangle$ is both $K(H_n W, n+1)$-null (since F is a GEM and $F\langle n\rangle$ is a retract of F) and $K(H_n W, n+1)$-supported. Therefore $F\langle n\rangle \simeq *$ and F has a single homotopy group in dimension n.

C.8 Corollary: *If W is n-supported, and $F \to E \to X$ a fibration sequence over a connected space X and W is an n-supported space, then the error term* $\mathrm{Fib}(\mathbf{P}_{\Sigma W} F \to \mathrm{Fib}(\mathbf{P}_{\Sigma W} E \to \mathbf{P}_{\Sigma W} X))$ *has a single homotopy group in dimension* $\mathrm{conn} H_* W + 1$.

8. v_1-PERIODIC SPACES AND K-THEORY

Introduction

In this chapter we concentrate on the relationship between two localization functors: homological localization with respect to K-theory and homotopy localization with respect to the well known Adams map from a certain suspension of a mod-p Moore space to the Moore space. This map was constructed by Adams as a K-isomorphism from some suspension of a Moore space to the Moore space in the stable range. Using [C-N], [Oka], [M-R] we know that one may choose the minimal dimensions of the Moore space at the range of the Adams map to be 3 for p odd and 5 for $p = 2$. Thus the Adams maps are $\Sigma^\ell v_1 : \Sigma^{q+\ell} M^3(\mathbb{Z}/p\mathbb{Z}) \longrightarrow \Sigma^\ell M^3(\mathbb{Z}/p\mathbb{Z})$, where $\ell \geq 0$, $q = 2p - 2$ for p odd, while for $p = 2$ the lowest dimensional map is $v_1 : M^{13}(\mathbb{Z}/p\mathbb{Z}) \longrightarrow M^5(\mathbb{Z}/p\mathbb{Z})$, i.e. the number of suspensions is $q = 8$ for $p = 2$.

It turns out that, as envisaged by Mahowald, Miller, Ravenel and others and proved in [M-T], [T], [B-4] under mild assumptions, the two are closely related and, moreover, the mod-p homotopy groups of these localizations $\pi_*(\mathbf{L}_K X, \mathbb{Z}/p\mathbb{Z})$ and $\pi_*(\mathbf{P}_{v_1} X, \mathbb{Z}/p\mathbb{Z})$ (where $\mathbf{P}_{v_1}$ as usual is the nullification with respect to the mapping cone of v_1) are related to $\pi_*(X, \mathbb{Z}/p\mathbb{Z})$ by a simple algebraic localization of the latter with respect to first-order operation of v_1 by composition. As we saw above homological localization is always a special case of homotopical localization with respect to a certain map f between 'large' spaces. In general, the map f is quite inaccessible since we must take the wedge of all possible homology isomorphisms among two spaces with cardinality not bigger than a well-chosen cardinality λ. This λ can be taken as the first infinite cardinal with $|E_*(pt)| \leq \lambda$.

However, in practice we now know that under quite mild assumptions the above map f can be taken to be a well known map, such as the Adams map for mod-p K-theory, or a map between a wedge of circles for integral homology [DF-4].

Thus, *loosely* speaking, the homotopy groups of $\mathbf{L}_K X$ should be, under mild assumptions, isomorphic to the algebraic localization of $\pi_*(X, \mathbb{Z}/p\mathbb{Z})$ with respect to v_1. This allows one to prove several results that are in close analogy with the classical results about rationalization. In this chapter we give an exposition of this analogy. Still, some severe limitations remain. Firstly we know that v_1 localization renders spaces periodic and, in particular, their mod-p homotopy groups become periodic after this localization process. But this also means that 'periodic families' which in the original space started in an arbitrarily high dimension are presumably 'pulled down' to the lower dimension within the first range of periodicity. This means, in particular, that even if one starts with a highly connected space of finite type, its v_1-localization will have, in general, low connectivity and will have an infinite number of v_1-periodic families starting at the first few dimensions. Hence this periodic space is expected to be 'thick' and not of finite type. Nonetheless its K-theory is the same as that of the original space, since we construct it by gluing on K-isomorphisms.

We shall not treat here localizations with respect to higher v_n-maps, but Bousfield shows that the initial steps are similar although the deeper properties seem much harder [B-4]. A major obstruction to the natural generalization is the failure of the telescope conjecture of Ravenel. This failure has prevented up to now the characterization, in the stable category, of $K(n)$-local infinite loop spaces in algebraic terms of the homotopy groups, where $K(n)$ denotes as usual the Morava K-theory.

In order to demonstrate the structures that we seek, we first consider the classical case of localization with respect to (a subring of) the ring of rational numbers $\mathbb{Q}$.

Rational homology and rational homotopy: Let us summarize some well-known results, due mostly to Serre and Quillen, about the relationship between rational homotopy and homology. This will serve us mostly as a paradigm for a similar relationship between 'periodic v_1-homotopy' and K-theory. For each prime $p \geq 2$ we have an operation on $\pi_* X$: multiplication by p, $- \times p : \pi_n X \to \pi_n X$. Tensoring with rationals is the universal manner of turning all these maps into isomorphisms.

For 1-connected (or even nilpotent) space one can associate with this operation of tensoring a space level rational localization $X \longrightarrow X_{\mathbb{Q}}$ that relates nicely to the algebraic operation of tensoring with the rationals $\mathbb{Q}$, compare e.g. (1.E.2-3). Let X, Y be simply connected spaces:

(1) A map $f : X \to Y$ induces an isomorphism on $\pi_* \otimes \mathbb{Q}$ if and only if it induces an isomorphism on rational homology.

(2) The homotopy groups of $X_{\mathbb{Q}}$, the rationalization of X, are canonically isomorphic to $\pi_* X \otimes \mathbb{Q}$.

(3) The map $X \to X_{\mathbb{Q}}$ induces an isomorphism on the rational homology groups.

(4) The Quillen rationalization functor $X \to X_{\mathbb{Q}}$ is naturally equivalent to the homotopy localization $\mathbf{L}_{\bigvee p}$ at the map $\bigvee p : \bigvee_p S^2 \to \bigvee_p S^2$ of the infinite wedge of 2-spheres into itself.

(5) The class of nilpotent, *rationally acyclic* spaces, i.e. $\{W | \tilde{H}_*(W, \mathbb{Q}) \simeq 0\}$, is precisely the class of spaces supported by $\bigvee_p M^3(\mathbb{Z}/p\mathbb{Z})$. Further, if we consider only simply connected spaces that are $\mathbb{Q}$-acyclic, then all such spaces are $\bigvee_p M^2(\mathbb{Z}/p\mathbb{Z})$-cellular. Thus the basic building block of $\mathbb{Q}$-acyclic 1-connected spaces are Moore spaces $M^2(\mathbb{Z}/p\mathbb{Z}) = S^1 \cup_p e^2$.

Proof: Of the above (1)–(3) are well-known facts and (4) is explained in (1.E.2) above. So we turn to (5). Compare [Bl-2 ,3.1]. Since $\tilde{H}_*(M^n(\mathbb{Z}/p\mathbb{Z}), \mathbb{Q}) \simeq 0$, every space supported by $M^n(\mathbb{Z}/p\mathbb{Z})$ for any prime p and any $n \geq 1$ is rationally acyclic. Given a space W with $\tilde{H}_*(W, \mathbb{Q}) \cong 0$ and $\pi_1 W \simeq 0$, we claim it is supported by $T = \bigvee_p M^3(\mathbb{Z}/p\mathbb{Z}) = \bigvee_p M(\mathbb{Z}/p\mathbb{Z}, 2)$ (i.e. it has non-trivial reduces integral homology only in dimension 2). To see this, notice that $W_{\mathbb{Q}} \simeq *$ by (3). We claim that $\mathbf{P}_T W \simeq *$. This is so, since $\mathbf{P}_T W$ being T-null has no p-torsion for any p and its rationalization $(\mathbf{P}_T W)_{\mathbb{Q}} \simeq W_{\mathbb{Q}} \simeq *$. The latter follows using e.g. (1.D.3) to the actual construction of $\mathbf{P}_T$ as a homotopy colimit, at each step the $\mathbb{Q}$- localization is contractible so it is also contractible when applied to the homotopy colimit. Thus our space $\mathbf{P}_T W$ has neither torsion nor rational homotopy and so it is contractible as claimed: $\mathbf{P}_T W \simeq *$ or $T < W$, as claimed. Further, by (3.B.3), since $T = \Sigma(\bigvee_p M^2(\mathbb{Z}/p\mathbb{Z}))$, we deduce $\bigvee_p M^2(\mathbb{Z}/p\mathbb{Z}) \ll W$, as claimed.

A. The v_1-periodization of spaces

Given the Adams map $v_1 : \Sigma^q M^3(\mathbb{Z}/p\mathbb{Z}) \to M^3(\mathbb{Z}/p\mathbb{Z})$, the most direct way to proceed in inverting v_1 is to consider the localization functor $\mathbf{L}_{v_1}$ as in Chapter 1 above. This functor will turn (in a functorial fashion) any space X into a v_1-periodic space. In particular, the mod-p homotopy groups: $\pi_n(\mathbf{L}_{v_1} X, \mathbb{Z}/p\mathbb{Z}) = [M^n(\mathbb{Z}/p\mathbb{Z}), \mathbf{L}_{v_1} X]$, are periodic with period q. This periodicity isomorphism is obtained by composing a given element $\alpha : M^n(\mathbb{Z}/p\mathbb{Z}) \to X$ of $\pi_n(X, \mathbb{Z}/p\mathbb{Z})$ with the Adams map which, as Adams proved, is a K-isomorphism $v_1 : M^{n+q}(\mathbb{Z}/p\mathbb{Z}) \to M^n(\mathbb{Z}/p\mathbb{Z})$ to get $v_1 \cdot \alpha \in \pi_{n+q}(X, \mathbb{Z}/p\mathbb{Z})$. We can consider the graded polynomial ring $\mathbb{Z}/p\mathbb{Z}[v_1]$ with $\deg v_1 = q$ as operating on $\pi_*(X, \mathbb{Z}, p\mathbb{Z})$. By inverting v_1 we can localize the v_1-module $\pi_*(X, \mathbb{Z}/p\mathbb{Z})$ to get a v_1-local module $v_1^{-1}\pi_*(X, \mathbb{Z}/p\mathbb{Z})$.

Explicitly, $v_1^{-1}\pi_*(X, \mathbb{Z}/p\mathbb{Z})$ is the direct limit of

$$\pi_*(X, \mathbb{Z}/p\mathbb{Z}) \to \pi_{*+q}(X, \mathbb{Z}/p\mathbb{Z}) \to \pi_{*+2q}(X, \mathbb{Z}/p\mathbb{Z}) \to \cdots.$$

However, we saw above that localization $\mathbf{L}_f$ behaves much better for the case $f : \Sigma W \to *$. Therefore, we try to consider localization (nullification) with respect to the mapping cone of v_1, or rather its suspension (see Chapter 3). So let $\Sigma(M^3(\mathbb{Z}/p\mathbb{Z}) \cup_{v_1} CM^{3+q}(\mathbb{Z}/p\mathbb{Z})) = \Sigma V(1)$.

Here, as usual, $V(1)$ denotes the mapping cone of the Adams map v_1. Since v_1 induces an isomorphism on K-theory, the space $V(1)$ is K-acyclic.

It is clear from the discussion in section 5.D above that localization with respect to $\Sigma V(1)$, i.e. the functor $\mathbf{P}_{\Sigma V(1)}$, is 'weaker' than $\mathbf{L}_{v_1}$; in other words $\mathbf{L}_{v_1}\mathbf{P}_{\Sigma v_1} \simeq \mathbf{P}_{\Sigma v_1}\mathbf{L}_{v_1} \simeq \mathbf{L}_{v_1}$. But for any X the localization $\mathbf{P}_{\Sigma v_1} X$ has periodic homotopy groups above the first dimension, since it is always local with respect to $\Sigma^2 v_1 : \Sigma^2\Sigma^q M^3(\mathbb{Z}/p\mathbb{Z}) \to \Sigma^2 M^3(\mathbb{Z}/p\mathbb{Z})$.

This will force us to restrict ourselves to a double loop space $\Omega^2 X$ in order to get good control. According to (5.D.1–2) any $\Sigma V(1)$-null space is also Σv_1-local space. In general, we do not know the relation between $\mathbf{L}_{v_1}$ and $\mathbf{P}_{\Sigma V(1)}$ since v_1 is not a suspension map. Notice also that the general localization functor $\mathbf{L}_p$ with respect to the degree-p map $p : S^1 \to S^1$ on the circle is much more complicated than the localization with respect to its suspension $\mathbf{L}_{\Sigma p}$ with $\Sigma p : S^2 \to S^2$, when applied to non-nilpotent space [Ca]. However, the relations between $\mathbf{L}_{\Sigma^k v_1}$ and $\mathbf{P}_{\Sigma^\ell V(1)}$ for various k, ℓ are not difficult to consider in view of Chapter 5.

Without being the 'full' v_1-localization $\mathbf{P}_{\Sigma V(1)} X$ still has v_1-periodic homotopy groups above the bottom dimension:

A.1 PROPOSITION: *The algebraic localization map $\pi_* \mathbf{P}_{\Sigma V(1)} X \to v_1^{-1} \pi_i \mathbf{P}_{\Sigma V(1)} X$ is an epimorphism for $i = 3$ and an isomorphism for $i > 3$.*

Proof: This is immediate from the usual function-space fibration sequence associated with the cofibration that defines $V(1)$.

A.2 Realizing $v_1^{-1}\pi_*(X, \mathbb{Z}/p\mathbb{Z})$ as a mod-p homotopy of a mapping telescope
One way to invert the action of v_1 on the mod-p homotopy of a given space is to look for a direct way to build a space whose mod-p homotopy groups are precisely the algebraic v_1-localization of the homotopy groups of the given space. This cannot be done in general by an idempotent coaugmented functor, but there is a nice and useful telescope that realizes these groups in a canonical fashion. Consider, after [M-T], the following analogue of the direct limit construction of a rationalization functor. We shall soon see (A.7) that this infinite telescope realizes as a space the v_1-periodic homotopy groups $v_1^{-1}\pi_*(X, \mathbb{Z}/p\mathbb{Z})$ above.

A.3 DEFINITION: *Let $\omega : \Sigma^q W \to W$ be any 'self-map' from some suspension of W to W. Define $T_\omega X$ to be the homotopy direct limit (i.e. infinite mapping telescope) of the tower of function complexes:*

$$\mathrm{map}_*(W, X) \xrightarrow{\omega^*} \mathrm{map}_*(\Sigma^q W, X) \xrightarrow{(\Sigma^q \omega)^*} \mathrm{map}_*(\Sigma^{2q} W, X) \to \cdots.$$

An important property of $T_\omega X$ is given by:

A.4 PROPOSITION: *For a map of finite spaces $\omega : \Sigma^k W \to W$, the telescope $T_\omega X$ is naturally an infinite loop space.*

Proof: Mapping a finite space to a direct limit of a telescope commutes with taking factors through the tower. Thus clearly we have a homotopy equivalence $\Omega^k T_\omega X \simeq T_\omega X$, making $T_\omega X$ into a periodic infinite loop space. This periodicity equivalence is induced on $T_\omega X$ by the map $\omega^* : (X^{\Sigma^n W})^W \to (X^{\Sigma^n W})^{\Sigma^k W} \simeq$

$(\Omega^k X^{\Sigma^k W})^W \simeq (X^{\Sigma^{n+k} W})^W$. The composition $(X^{\Sigma^n W})^W \to (X^{\Sigma^{n+k} W})^W$ is simply $(\omega^*)^W$.

Further, one can conclude from the above argument:

A.5 PROPOSITION: *For any map of finite complexes $\omega : \Sigma^q W \to W$, the space $T_\omega X$ is ω-local.*

Proof: We have $\mathrm{map}_*(W, T_\omega X) \simeq \mathrm{map}_*(W, \Omega^q T_\omega X) \simeq \mathrm{map}_*(\Sigma^q W, T_\omega X)$ and this composition is induced by ω as above. In more detail: The map in question, ω^* : $\mathrm{map}_*(W, T_\omega X) \to \mathrm{map}(\Sigma^q W, T_\omega X)$, is the direct limit of a tower of maps since W is a finite complex:

$$\omega^* : \mathrm{map}_*(W, X^{\Sigma^{\ell q} W}) \to \mathrm{map}_*(\Sigma^q W, X^{\Sigma^{\ell q} W})$$

which is the same as s^W where s is

$$s : \mathrm{map}_*(\Sigma^{\ell q} W, X) \to \mathrm{map}_*(\Sigma^{(\ell+1)q} W, X).$$

Since the shift map of towers induces a homotopy equivalence on their common limit $T_\omega X$, it is still an equivalence after taking $\mathrm{map}_*(W, s)$.

A.6 PROPOSITION: *If $F \to E \to X$ is a fibration sequence in $\mathcal{S}_*$, then so is $T_\omega F \to T_\omega E \to T_\omega X$, where the homotopy fibre is taken over the null component of the telescope.*

Proof: Both function complex functor $\mathrm{map}_*(W, -)$ and telescopes preserve fibrations see Appendix HL.

THE MAHOWALD-THOMPSON TELESCOPE OF THE ADAMS MAP v_1 Let us consider the mod-p homotopy groups of this infinite telescope. For any finite complex V there is a natural homotopy equivalence, which is the map induced on the homotopy colimit:

$$\operatorname*{hocolim}_k \mathrm{map}_*(V, \mathrm{map}_*(\Sigma^{kq} W, X)) \simeq \mathrm{map}_*(V, T_\omega X).$$

In particular, for the path component level we have a direct limit of sets of components:

$$\mathrm{dirlim}[V, X^{\Sigma^{kq} W}] \simeq [V, T_\omega X].$$

We denote by $\pi_k(X, V)$ the homotopy classes $[\Sigma^k V, X]$. Thus $\pi_k(X, M^3(\mathbb{Z}/p\mathbb{Z})) \simeq [M^{k+3}, X] \simeq \pi_{k+3}(X, \mathbb{Z}/p\mathbb{Z})$. Notice the shift in dimension with respect to the usual notation of homotopy groups with coefficient in $\mathbb{Z}/p\mathbb{Z}$.

It will be more convenient from the point of view of indexing to use $\pi_*(X, M)$ with $M = M^3(\mathbb{Z}/p\mathbb{Z})$ rather than $\pi_*(X, \mathbb{Z}/p\mathbb{Z})$.

We now turn to special properties of T_ω, where $\omega = v_1$ is the Adams map that induces an isomorphism on K-theory. In this case we denote the telescope by T_{v_1}. In particular, we have from the above considerations that the telescope T_{v_1} realizes the v_1-periodic homotopy:

We have defined algebraically $v_1^{-1}\pi_*(X, M)$ as the localization of the groups $\pi_*(X, M)$ with respect to the composition operation by $v_1\colon \Sigma^q M \to M$.

A.7 PROPOSITION: *There is a canonical isomorphism* $v_1^{-1}\pi_*(X, M) \simeq \pi_* T_{v_1} X$.

Proof: We have $\pi_* T_{v_1} X = \operatorname{colim} \pi_* \operatorname{map}_*(\Sigma^{q\ell} M, X)$. ∎

In light of this canonical identification we will use $T_{v_1} X$ instead of $v_1^{-1}\pi_* X$. The former has the advantage of being a space.

Not only the homotopy groups of the telescope are periodic, the space itself is local with respect to K-theory [T]:

A.8 PROPOSITION: $T_{v_1} X$ *is always* K*-local in the sense of Bousfield.*

Proof: In [B-3] Bousfield proves that an infinite loop space is K-local if and only if its mod-p homotopy groups are v_1-periodic.

RELATIONS BETWEEN K-HOMOLOGICAL LOCALIZATION, THE NULLIFICATION $\mathbf{P}_{v_1}$ AND THE TELESCOPE. We now use the above general construction to show that a small modification of both the nullification functor ('periodization' might be a better name here) and the K-homological localization are closely related to the much more accessible telescope $T_{v_1} X$.

Start by specializing the above to $W \in \mathcal{S}_*^p$, which is also finite dimensional and n-supported (7.C):

A.9 PROPOSITION [B-4, 11.5]: *Let* $\omega\colon \Sigma^k W \to W$ *be a self-map of a finite space* W *with* $C = \operatorname{cof}(\omega)$*. For any* $X \in \mathcal{S}_*$ *the map* $j : X \to \mathbf{P}_{\Sigma C} X$ *induces an equivalence* $T_\omega(j) : T_\omega X \to T_\omega(\mathbf{P}_{\Sigma C} X)$.

Proof: By (5.C.5) above, since C is a finite space the homotopy fibre of is a finite Postnikov stage—namely all its higher homotopy groups vanish above some fixed dimension d=dim $C + 1$. Therefore, taking telescopes we get the following infinite ladder of pointed function complexes and their nullifications:

$$\begin{array}{ccccccc}
\mathbf{P}_{\Sigma C}(X^W) & \xrightarrow{\omega^*} & \mathbf{P}_{\Sigma C}(X^{\Sigma^q W}) & \xrightarrow{(\Sigma^q\omega)^*} & \mathbf{P}_{\Sigma C}(X^{\Sigma^{2q}W})\cdots & \longrightarrow & \mathbf{P}_{\Sigma C}T_\omega X \\
\downarrow & & \downarrow & & \downarrow & & \downarrow \\
(\mathbf{P}_{\Sigma C}X)^W & \xrightarrow{\omega^*} & (\mathbf{P}_{\Sigma C}X)^{\Sigma^q W} & \xrightarrow{(\Sigma^q\omega)^*} & (\mathbf{P}_{\Sigma C}X)^{\Sigma^{2q}W}\cdots & \longrightarrow & T_\omega \mathbf{P}_{\Sigma C}X
\end{array}$$

using [HL], the fact that $\mathbf{P}_A$ commutes with telescopes for a finite space A (1.D.6), and that taking homotopy fibre commutes with taking telescopes we deduce that the homotopy fibre of $T_\omega X \simeq \mathbf{P}_{\Sigma C}T_\omega X \to T_\omega \mathbf{P}_{\Sigma C}X$ is also a finite Postnikov stage. But by periodicity of $T_\omega Y$, that fibre is a fibre of maps between two periodic infinite loop spaces (A.4) so the homotopy fibre must be contractible as needed. The first equivalence in the last map comes from $C = \mathrm{cof}(\omega)$. ∎

REMARK: We now come to the main formula of this chapter. Notice that $T_\omega X$ and $\mathbf{P}_{\Sigma C}$ whose $C = \mathrm{cof}(\omega)$ are closely related: Both associate a $\Sigma^2\omega$-local space to any space X, both 'come close' to preserving fibrations. Notice also that $(\mathbf{P}_{\Sigma C}X)^{\Sigma^2 W}$ is an infinite loop space, since we have a cofibration $\Sigma C \to \Sigma^2\Sigma^q W \to \Sigma^2 W$, hence $\mathrm{map}_*(\Sigma^2 W, \mathbf{P}_{\Sigma C}X) \simeq \mathrm{map}_*(\Sigma^q\Sigma^2 W, \mathbf{P}_{\Sigma C}X)$ or $(\mathbf{P}_{\Sigma C}X)^{\Sigma^2 W} \simeq \Omega^q(\mathbf{P}_{\Sigma C}X)^{\Sigma^2 W}$.

The next proposition shows a relationship between $(\mathbf{P}_C X)^W$ and $T_\omega X$.

We formulate the main reduction of the nullification functor with respect to the cone of a self-map to the telescope of this map. This result, when applied to the only non-nilpotent self-maps on finite p-torsion space, yields a better understanding of the K-localization and periodization of a space. The proof relies heavily on the main results of this and previous chapters. The proposition says that, while the localization itself might still be mysterious, if one looks at the mod-p information or the function complex of maps from the relevant Moore space to the localization, it is as simple as one can expect—after double looping: We first formulate a general result about self-maps of finite complexes that, inter alia, shows how to modify $\mathbf{P}_C$ so that it preserves fibration sequences unconditionally.

A.10 PROPOSITION: *If $C = \mathrm{cof}(\Sigma^q W \xrightarrow{\omega} W)$ is the mapping cone of a self-map of a finite space W, then there is a homotopy equivalence*

$$\Omega T_\omega \Omega X \simeq \Omega(\mathbf{P}_C \Omega X)^W$$

or equivalently,

$$\Omega^2 T_\omega X \simeq \Omega^2(\mathbf{P}_{\Sigma C}X)^W.$$

Proof: By the definition of C as a cofibre, we have a cofibration

$$\Sigma C \to \Sigma^2\Sigma^q W \to \Sigma^2 W.$$

This shows that $(\mathbf{P}_{\Sigma C}X)^{\Sigma^2 W} \simeq (\mathbf{P}_{\Sigma C}X)^{\Sigma^{q+2}W}$ is a homotopy equivalence, since this induced map is a part of a fibration with contractible base. Furthermore, this can be continued to a telescope of equivalences $\{(\mathbf{P}_{\Sigma C}X)^{\Sigma^2\Sigma^{q\ell}W}\}_{\ell\geq 0}$ (where we have suppressed the maps from the notation) that converges by the above definition of T_ω as a functor, to $\Omega^2 T_\omega \mathbf{P}_{\Sigma C}X$.

Therefore we get an equivalence

$$(\mathbf{P}_{\Sigma C}X)^{\Sigma^2 W} \stackrel{\simeq}{\to} \Omega^2 T_\omega \mathbf{P}_{\Sigma C}X.$$

Now using A.9 above one can rewrite the right-hand side:

$$\Omega^2 T_\omega \mathbf{P}_{\Sigma C}X \simeq \Omega^2 T_\omega X.$$

Use (3.A.1) to obtain an equivalence:

$$\Omega^2(\mathbf{P}_{\Sigma C}X)^W = \Omega(\mathbf{P}_C\Omega X)^W \simeq \Omega T_\omega \Omega X \simeq \Omega^2 T_\omega X,$$

as needed. ∎

A.11 COROLLARY: *If $C = \mathrm{cof}(\omega)$ is a finite space then the (non-augmented) functor that takes X to $(\mathbf{P}_C\Omega X)^{\Sigma W} \simeq \Omega^2(\mathbf{P}_{\Sigma C}X)^W$ preserves fibre sequences.*

Proof: Both T_ω and Ω preserve fibre sequences.

A.11.1 EXAMPLES: First a trivial example. If $W = S^n$ and our map $\omega = \Sigma W \to W$ is just the null map $S^{n+1} \to S^n$ with cofibre $S^n \vee S^{n+2}$, then $\mathbf{P}_C = \mathbf{P}_\omega$ is $\mathbf{P}_{S^n}$, i.e. the $(n-1)$-Postnikov stage. $\mathbf{P}_{\Sigma C}$ is $\mathbf{P}_n$, the nth Postnikov stage, and $\Omega^2(\mathbf{P}_n X)^{S^n} = \Omega^{n+2}\mathbf{P}_n X \simeq *$. This is the trivial functor, so it certainly preserves fibrations. This example underlines the radical modification that is sometimes done here to a functor like the Postnikov section functor that makes it preserve fibrations unconditionally.

For a more interesting example consider in (A.10) the Adams self-map $\omega = v_1$. In this case the function complex $\mathrm{map}_*(M, \mathbf{P}_{\Sigma C}X)$, where $C = V(1)$, the mapping cone of v_1, contains in fact the most interesting information about the nullification functor, namely its mod-p homotopy groups. So in this case Proposition

A.10 precisely identifies the relevant homotopy groups of the up-to-now mysterious $\Sigma V(1)$-nullification functor.

The last result shows that $T_\omega X$ is very closely related to the space of maps $\mathrm{map}_*(W, \mathbf{P}_\omega X)$ of nullification with respect to $\mathrm{cof}(\omega)$. Further, we now deduce the same relationship to Bousfield homological localization. This establishes a relation between v_1-periodic homotopy and K-theory.

Proposition A.10 has an immediate corollary relating the telescope, nullification and K-homological localization. It shows in essence that in the category of p-torsion spaces modulo low dimensions, and after concentrating on the mod-p information, it is sufficient to homotopically invert one K-isomorphism, the Adams map, in order to invert all K-isomorphisms.

A.12 THEOREM: *For any $X \in \mathcal{S}^p_*$ with p an odd prime, we have homotopy equivalences*

$$\Omega^2 T_{v_1} X \simeq \Omega(\mathbf{P}_{v_1}\Omega X)^M \to \Omega(\mathbf{L}_K \Omega X)^M$$

where $M \simeq M^3(\mathbb{Z}/p\mathbb{Z})$.

Outline of Proof: This follows from (A.8) above when coupled with (A.10), since by (A.10) the function complex into the nullification is K-local.

B. K-isomorphisms, K-acyclic spaces

First we begin with a theorem of Thompson:

B.1 THEOREM: *Let X be a 3-connected space. If $\tilde{K}_*(\Omega^3 X) \simeq 0$ then $T_{v_1} X \simeq *$.*

Proof: Since the space $M = S^2 \cup_p e^3$ is three dimensional we have a $M \to S^3 \to S^3$ cofibration that gives a principal fibration $\Omega^3 X \to \Omega^3 X \to X^M$ whose base is connected, since $X \in \mathcal{S}_*\langle 3\rangle$. Therefore by (1.H.2), $\tilde{K}_*(X^M) \simeq 0$. This means that any map $X^M \to T_{v_1} X$ is null homotopic, since $T_{v_1} X$ is a K-local space and the said map factors through $\mathbf{L}_K X \simeq *$. Now take the natural maps of the telescope that defines $T_{v_1} X$ to $T_{v_1} X$ itself, namely

$\alpha_0 : \mathrm{map}_*(M, X) \to T_{v_1} X,$

$\alpha_\ell : \mathrm{map}_*(\Sigma^{q\ell} M, X) \to T_{v_1} X.$

Since by definition $\Omega^q \alpha_\ell = \alpha_{\ell+1}$ we see that $\alpha_1 = \Omega^q \alpha_0 \simeq *$ and $\alpha_\ell \simeq *$, and this immediately implies $T_{v_1} X \simeq *$, since $\mathrm{colim}(X_i \to X_\infty)$ is the identity on $X_\infty = Tel(X_i)$ for any telescope $X_i \to X_{i+1} \cdots$ ∎

SPACES WITH TRIVIAL NULLIFICATION $\mathbf{P}_{\Sigma V(1)}$: Recall that $M = M^3(\mathbb{Z}/p\mathbb{Z})$ is the 3-dimensional Moore space and $V(1) = \mathrm{cof}(v_1 : \Sigma^q M \to M)$. The space $V(1)$ is clearly 2-supported, see (7.C.7). Although the relation between the nullification and the telescope goes in (A.10) through function complexes, in the case of a trivial telescope or nullification the relation is more direct.

B.2 THEOREM: *If X be an n-supported M-cellular space with n=conn$X$+1 then* $\mathbf{P}_{\Sigma V(1)}\Sigma X \simeq *$ *if and only if there exists an integer $N \geq 0$ with $T_{v_1}\Sigma^N X \simeq *$.*

REMARK: The conditions are satisfied by any finite p-torsion 2-connected space [Bl-2, 3.1].

Proof: Thus we must show that $\mathbf{P}_{\Sigma V(1)}$ kills the suspension of X iff the v_1-telescope kills some, possibly higher, suspension of X. We will prove one direction; the other direction follows similarly by reversing the arguments. Assume $T_{v_1}\Sigma^N X \simeq *$. First, since the integer q is at least 2, we have as in Theorem A.10 and (A.11.1) above the homotopy equivalence

$$\left(\mathbf{P}_{\Sigma V(1)}X\right)^{\Sigma^q M} \simeq T_{v_1}\mathbf{P}_{\Sigma V(1)}X.$$

From the assumption we get by Proposition A.9 above $T_{v_1}\mathbf{P}_{\Sigma V(1)}\Sigma^N X \simeq *$. But, as earlier, we get from $q \geq 2$ that

$$(\mathbf{P}_{\Sigma V(1)}\Sigma^N X)^{\Sigma^q M} \simeq T_{v_1}\mathbf{P}_{\Sigma V(1)}\Sigma^N X$$

and so the left-hand side $\simeq *$. Using the cofibration $\Sigma\Sigma^q M \to \Sigma M \to \Sigma V(1)$ that induces a fibration upon taking a function complex into $\mathbf{P}_{\Sigma V(1)}\Sigma^N X$, we conclude that $\mathrm{map}_*(\Sigma M, \mathbf{P}_{\Sigma V(1)}\Sigma^N X) \simeq *$, since it is the total space in a fibration with contractible fibre and base ($\mathbf{P}_{\Sigma V(1)}\Sigma^N X$ is by definition $\Sigma V(1)$-null). But for $N \geq 1$ the space $\mathbf{P}_{\Sigma V(1)}\Sigma^N X$ is ΣM-cellular by the construction of $\mathbf{P}_{\Sigma V(1)}$ as a homotopy limit. Therefore, we can conclude from the Whitehead theorem (2.E.1) that $\mathbf{P}_{\Sigma V(1)}\Sigma^N X \simeq *$. If $N \in \{0,1\}$ we are done; otherwise, since X is n-supported we have by (7.B.4) :

$$\langle \Sigma X \rangle = \langle \Sigma^N X \rangle \vee K(\mathbb{Z}/p\mathbb{Z}, c+1),$$

where c = connΣX + 1. But by discussion in example 3.C.5 we know that $\mathbf{P}_{\Sigma V(1)}K(\mathbb{Z}/p\mathbb{Z}, 3) \simeq *$ and conn$X \geq 1$, so we can conclude that $\mathbf{P}_{\Sigma V(1)}\Sigma X \simeq *$, as needed. The other direction is obtained more easily by reversing the above arguments. ∎

We now apply (B.2) to K-theory.

In the next few results we relate the vanishing of K-theory on some iterated loop space of a given space X to the partial ordering ($<$) and ($\ll$). The idea here is to have a K-theoretic analog of item (5) in section A above concerning trivial rational homotopy. We see in (B.6) a reasonable analog to item (5) in the Introduction to the present section.

B.3 COROLLARY: *If* $\tilde{K}_*\Omega^3 X \simeq 0$ *then* $\mathbf{P}_{\Sigma W(1)}\Sigma X \simeq *$.

Proof: $T_{v_1}X \simeq *$ by Proposition B.1 above, so by Theorem B.2, $\mathbf{P}_{\Sigma V(1)}\Sigma X \simeq *$.

B.4 COROLLARY: *If* $\tilde{K}_*X \simeq 0$ *then* $\mathbf{P}_{\Sigma V(1)}\Sigma^3 X \simeq *$.

Proof: This is true since we have $\tilde{K}_*(\Omega^3\Sigma^3 X) \simeq 0$, so $\mathbf{P}_{\Sigma V(1)}\Sigma^3\Omega^3\Sigma^3 X \simeq *$. But $\Sigma^3\Omega^3 Y \ll Y$. Take $Y = \Sigma^3 X$.

B.5 COROLLARY: *If* $\tilde{K}_*(X) \simeq 0$ *and* X *is 1-connected then* $\mathbf{P}_{\Sigma V(1)}\Sigma X \simeq *$.

Proof: $\langle \Sigma X \rangle = \langle \Sigma^3 X \rangle \vee K(\mathbb{Z}/p\mathbb{Z}, \mathrm{conn}\Sigma X + 1)$ and $\mathbf{P}_{\Sigma V(1)}K(\mathbb{Z}/p\mathbb{Z}, 3) \simeq *$, while $\mathrm{conn}\Sigma X + 1 \geq 3$.

B.6 COROLLARY:

$$\begin{aligned} \tilde{K}_*(\Omega X) \simeq * &\Rightarrow \mathbf{P}_{\Sigma V(1)}X \simeq * \\ &\Rightarrow V(1) \ll X. \end{aligned}$$

Proof: In general $\Sigma\Omega X \ll X$ (3.C.7). Now by (1.A.8) the assumption implies $\mathbf{P}_{\Sigma V(1)}\Sigma\Omega X \simeq *$. Therefore $\mathbf{P}_{\Sigma V(1)}X \simeq *$. But then the conclusion follows by (3.B.3). ∎

9. CELLULAR INEQUALITIES

A. Introduction and main results

In previous chapters we have profitably used *cellular inequalities* of the form $W \ll X$ several times. Thus to show that $\mathbf{L}_f(\mathrm{GEM})$ is again a GEM, we used the observation $X \ll SP^k X$ for all $0 \le k \le \infty$, namely that $SP^k X$ is X-cellular. Also, in proving a cellular form of Bousfield's *key lemma* (4.D) above we show that $\Sigma^2 X \ll \Sigma(SP^k X/X)$. This last inequality can be rephrased by saying that ΣX is the *leading term* in the construction of $\Sigma SP^k X$ out of ΣX and all *higher-order* terms are $\Sigma^2 X$-cellular.

We observe that if $W \ll X$, namely if X can be built from W by a possibly infinite process of taking cofibration or iterated pointed homotopy colimits, then X retains several important properties of W. For example, $\mathrm{conn} W \le \mathrm{conn} X$. Another example is the vanishing of generalized homology or the vanishing of the singular homology with coefficients in G up to some fixed dimension d. If $W \ll X$, then any homology theory that vanishes on W also vanishes on X. Still another property is one associated with each given space T in $\mathcal{S}$. That property is the 'vanishing' of the pointed mapping space $\mathrm{map}_*(W,T)$. If X is W-cellular, then X also has the same property. In particular, any cohomology theory vanishing on W also vanishes on X.

Notice that the above two properties are not preserved under unpointed homotopy colimits: Any space whatsoever is homotopy equivalent to an unpointed homotopy colimit of a diagram of contractible spaces (Appendix HL).

Let us list several problems whose understanding relates to cellular inequalities:

A.1 For a given homology theory E_* such as Morava K-theory, an important problem is to decide when an E_*-acyclic space is a direct limit of its *finite* E_* acyclic subcomplexes. This is closely related to the question of whether or not the theory E_* is 'smashing': namely whether or not Bousfield's E_*-localization is simply the smash product with the E_*-local sphere. We approach this by asking whether there exists a finite space A_E which is a generator of $\mathcal{C}^{\cdot}(\text{acyclics})=$ the closed (under pointed hocolims) class of all E_*-acyclic spaces, or a generator for a large subclass thereof: Namely we look for a finite A_E with $\mathcal{C}^{\cdot}(A_E) = \mathcal{C}^{\cdot}$ (acyclics). For example, we saw in (8.B.6) that the mapping cone of the Adams map $V(1)$ generates a large class of K-acyclic spaces.

A.2 Another example is the general problem of relating (co)limits to homotopy (co)limits. We would like to show that in some way the 'difference' between colim and hocolim is 'smaller than expected'. For example, often, as we saw in Chapter 4, if $\underset{\sim}{X}$ is a *pointed* diagram in $C^{\cdot}(A)$ (i.e. consisting of A-cellular spaces) then the fibre and cofibre of the natural map

$$\mathrm{hocolim}_* \underset{\sim}{X} \to \mathrm{colim} \underset{\sim}{X}$$

are in $C^{\cdot}(\Sigma A)$ or even in $C^{\cdot}(\Sigma^2 A)$.

A.3 Both the fibre and the cofibre of a map $g : X \to Y$ can be considered as 'measures of the deviation' of g from homotopy equivalence. Yet few general relations between them are available. For example, assume that g is E_*-homology isomorphism with the special property that the fibre of g is E_*-acyclic. What can we deduce on $\mathrm{cof}(g) = Y \cup CX$ beyond the immediate consequence that it is E_*-acyclic? It turns out that $\Omega(Y \cup CX)$ is also E_*-acyclic under this condition. For a proof see (D.2)(ii) below.

A.4 Moreover, inequalities will help us in general to find relations between $\mathrm{hocolim}_*$ 3$\mathrm{holim}_i X_{ij}$ and $\mathrm{holim}_i\ \mathrm{hocolim}_* X_{ij}$ for a given doubly indexed diagram. In general, inverse limit and direct limit functors do not commute, but often they are related by cellular inequalities such as $\Sigma\Omega X \ll \Omega\Sigma X$.

A.5 Recall (Compare [B-4] [DF-5]: We say that A 'supports' X and denotes it by $A < X$ if $P_A X \simeq *$ or, equivalently, if for all Y one has the implication $\mathrm{map}_*(A, Y) \simeq * \Rightarrow \mathrm{map}_*(X, Y)$.

A.6 Proposition: *For all $A, X \in \mathcal{S}_*$ one has*

$$\text{(i)} \qquad A \ll X \Rightarrow A < X,$$
$$\text{(ii)} \qquad \Sigma A < X \Rightarrow A \ll X.$$

Proof: (i) We know that X is built from A by a process of taking $\mathrm{hocolim}_*$. One needs to check that if in a diagram $\underset{\sim}{A}$ one has $A < A_\alpha$ for all α, then $A < \mathrm{hocolim}_* \underset{\sim}{A}$. This follows directly from (1.D), explicitly: Let Y be any A-null space so that $\mathrm{map}_*(A, Y) \simeq *$. we must show that $\mathrm{map}_*(\mathrm{hocolim}_* \underset{\sim}{A}, Y) \simeq$. Notice $\mathrm{map}_*(\mathrm{hocolim}_* \underset{\sim}{A}, Y)$ is equivalent to $\mathrm{holim}\ \mathrm{map}_*(\underset{\sim}{A_\alpha}, Y)$ which is contractible because $\mathrm{holim} V_\alpha$ with $V_\alpha \simeq *$ is contractible.

(ii) This is Proposition 3.B.3 .

Therefore, in some cases, we get interesting relations of the form $A < X$ by first showing $A \ll X$ and vice versa.

A.7 NOTATIONS: Recall that for any pointed diagram X and any diagram Y, we denote the *pointed* homotopy colimits by $\int_I \underset{\sim}{X}$ and the unpointed one by $\oint_I \underset{\sim}{X}$, namely:

$$\int_I \underset{\sim}{X} \equiv \operatorname{hocolim}_{*I} \underset{\sim}{X}$$

and

$$\oint_I \underset{\sim}{Y} \equiv \operatorname{hocolim}_I \underset{\sim}{Y}$$

for $\underset{\sim}{X}: I \to \mathcal{S}_*$ and $\underset{\sim}{Y}: I \to \mathcal{S}$ are diagrams.

This notation is supposed to remind the reader that in going from $\oint$ to $\int$, we must collapse to a point the classifying space of the small category $|I|$—which in the simplest connected case is a circle. It should also remind the reader that homotopy colimits, both pointed and unpointed, formally share some of the axiomatic properties of the definite integrals of real valued or other functions. Thus they are homotopy additive: hocolim of a homotopy pushout is weakly homotopy equivalent to the pushout of the homotopy colimits. The pointed homotopy colimit of 'the point' (i.e. the diagram that assigns a single point to each object in I) is a point, while in the unpointed homotopy colimit the point serves as a 'unit' and not as a 'zero' and the *unpointed, free* homotopy colimit of a diagram of points is the nerve or classifying space of the underlying small category that serves here as the 'domain of integration'. It is also additive in the obvious sense with respect to the 'domain of integration'. This notation also serves to emphasize the distinction between homotopy colimits and homotopy limits. Although they are formally dual from a homotopical algebra point of view, they share only few properties and it might be better not to create a false analogy between them.

We also find this notation shorter, more convenient and visually appealing.

Main Results

We consider two main directions: one concerns the homotopy fibres of a map between colimits; the other relates strict colimits to homotopy colimits.

A.8 THEOREM: *Let I be a small category and A a pointed space. Let $\{E_\alpha \to B_\alpha\}$ be an I-diagram of maps in $\mathcal{S}$ over connected spaces B_α with pointed homotopy fibres F_α. Assume that $\mathcal{U}$ is a closed class such that $F_\alpha \in \mathcal{U}$ for all α.*

In that case also:

(WCS) $$\operatorname{Fib}\Big(\oint_I E_\alpha \to \oint_I B_\alpha\Big) \in \mathcal{U}.$$

In particular, if $A \ll F_\alpha$ for all α then $A \ll \mathrm{Fib}(\oint E_\alpha \to \oint B_\alpha)$.

A similar statement holds for diagrams in $\mathcal{S}_*$ and pointed hocolim$_*$.

Proof: See (C.1) below.

A.9 Remarks and Examples:

(i) In several important cases a much stronger version of the above is true, namely

$$\text{(CS)} \qquad \oint_\alpha F_\alpha \ll \mathrm{Fib}\Big(\oint_\alpha E_\alpha \to \oint_\alpha B_\alpha\Big).$$

This is true if, for example, the situation falls under the assumptions of V. Puppe's theorem (Appendix HL): e.g. all the maps in $(B_\alpha)_{\alpha \in I}$ are identity maps and $|I| \simeq *$ so that I is a contractible category.

For a concrete example consider the diagram:

$$\begin{array}{ccc} * & \longrightarrow & * \\ \uparrow & & \uparrow \\ * & \longrightarrow & X \\ \downarrow & & \downarrow \\ * & \longrightarrow & * \end{array}$$

The stronger inequality (CS) gives $\Sigma\Omega X \ll \Omega\Sigma X$, which is true for any connected X (see 3.A). The last case is a special case of the inequality: For any map $f\colon X \to Y$ the homotopy fibre of the suspension of f can be built from the suspension of the homotopy fibre of f. Compare [Ch-2],[DF-5]:

$$\Sigma\mathrm{Fib}(X \to Y) \ll \mathrm{Fib}(\Sigma X \to \Sigma Y).$$

(ii) We refer to the inequality in the theorem as *weak* Cauchy-Schwartz (WCS) and to the above (CS) as Cauchy-Schwartz, since they are reminiscent of the inequality $x^2 + y^2 \leq (x+y)^2$ for non-negative real numbers, the analogue of the product being limit and that of the sum being colimit.

(iii) Non-trivial examples of the (WCS):

(1) The inequality $\Sigma F \ll B/E$ for any fibre sequence $F \longrightarrow E \longrightarrow B$ over a connected B and

(2) the inequality $F \ll \mathrm{Fib}(E/F \longrightarrow B)$ for such a fibration.

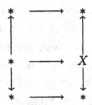

Proof: They arise in the maps of diagrams:

$$
\begin{array}{ccc}
* & \longrightarrow & * \\
\uparrow & & \uparrow \\
E & \longrightarrow & E \\
\downarrow & & \downarrow \\
E & \longrightarrow & B
\end{array}
\qquad \text{and} \qquad
\begin{array}{ccc}
* & \longrightarrow & * \\
\uparrow & & \uparrow \\
F & \longrightarrow & * \\
\downarrow & & \downarrow \\
E & \longrightarrow & B
\end{array}
$$

where the vertical arrows are maps inside diagrams and the horizontal ones are components of maps between diagrams and the map $E \to E$ is the identity.

So this is a special case of the above with I being $* \longleftarrow * \longrightarrow *$.

To derive the first inequality from (WCS) one uses the fact that for connected X, Y one has $\Sigma X \ll Y \Leftrightarrow X \ll \Omega Y$. See Chapter 3.

Consider the first diagram:

$$
\begin{array}{ccccc}
* & \longleftarrow & E & \longrightarrow & E \\
\downarrow & & \downarrow & & \downarrow \\
* & \longleftarrow & E & \longrightarrow & B
\end{array}
$$

The homotopy colimit of the top row is $(*)$ and of the bottom row is B/E. Then the map $\text{hocolim}_* F_\alpha \to \text{hocolim}_* B_\alpha$ is the map $* \to B/E$. Thus by Theorem A.8 above, $F \ll \Omega(B/E)$; by (3.A), $\Sigma F \ll B/E$. In fact, as we will see in (D.1) below one can express the cofibre B/E directly as a pointed homotopy colimit as follows: Recall (1.F), (HL) that the total space itself is the unpointed homotopy colimit of the diagram of 'fibres' indexed by ΓB. If we suspend this diagram of fibres at each object we get a pointed diagram—actually with two possible base points. We can then take the pointed homotopy colimit with respect to one of these two base points. One gets:

$$B/E \simeq \int_{\Gamma B} \Sigma F.$$

A striking example is:

A.10 COROLLARY: *The homotopy fibre of $X \to X/A$ is A cellular for any cofibration sequence $A \to X \to X/A$ in $\mathcal{S}_*$.*

Proof: The inverse image of a simplex in X/A is either a point or A itself, so by (A.11) below the homotopy fibre is A-cellular.

This corollary is clearly a special case of the more general theorem:

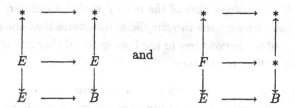

A.11 THEOREM: *Let $f: X \to Y$ be any map of connected pointed simplicial spaces. Assume that the inverse image $f^{-1}(\sigma)$ of each simplex σ in Y is connected and A-cellular. Then the homotopy fibre of the map f is also A-cellular. Similarly, if the map is a simplicial map between two simplicial complexes then one can consider the point inverses of all the barycentres in the base space; if they are all A-cellular then so is the homotopy fibre of the map.*

Proof: The proof is lengthy and will occupy the next section.

We will first demonstrate how to present the homotopy fibre as an unpointed homotopy colimit of a special diagram.

B. The homotopy fibre as homotopy colimit

We shall see below that it is often easier to present a space as an unpointed homotopy colimit of a specific kind than as a pointed one. This is clear in view of the fact that, given an I-diagram $\underset{\sim}{X}$, we cannot take its pointed hocolim unless the diagram is pointed. But there is no reason to assume that it has a consistent choice of base points, namely a map $pt \to \underset{\sim}{X}$ of I-diagrams. The latter is equivalent to $\lim_I \underset{\sim}{X} \neq \emptyset$. But we will be interested in cases when not only the limit of our diagram is empty, but also the *homotopy* inverse limit might be empty. The feature that renders an unpointed homotopy colimit non-trivial from the present point of view is the *restriction one puts on the homotopy type of the classifying space of I*. As we saw above it is otherwise pointless from the present point of view to consider the construction of spaces via unrestricted use of unpointed homotopy colimits. The restriction that occurs most naturally is that of restricting the possible shapes of the small categories over which one takes these unpointed homotopy colimits. This was investigated elsewhere in some detail [Ch-1,2]. For our purpose here we will consider in particular *contractible* small categories, namely categories I with $|I| \simeq *$. We will see immediately that unpointed homotopy colimits of a diagram of non-empty spaces X over a contractible small category I *can always be reconstructed up to a weak equivalence as a pointed homotopy colimit* out of the same spaces with a suitable choice of base points in these non-empty spaces X_α.

In applications, we often wish to obtain inequalities of the type $W \ll \operatorname{Fib}(X \to Y)$, since the fibre as opposed to cofibre is harder and often more useful to estimate.

It turns out that one way to arrive at such inequalities is to express the homotopy fibre $\operatorname{Fib}(X \xrightarrow{g} Y)$ of a map $g: X \to Y$ as a homotopy colimit of the inverse images of points $g^{-1}(y)$ for $y \in Y$. This is easier to consider when X, Y are simplicial sets, or in fact simplicial complexes. Let ΓY be the small category of all simplices of a simplicial complex Y with $|\Gamma Y| \simeq Y$ (see (1.F) above).

If Y is a simplicial complex, then ΓY is just the category of simplices of Y: It has one object for each simplex and one map for each face inclusion, so that $\mathrm{Hom}(\sigma, \sigma')$ in that category ΓY is either empty or a singleton.

We associate with the map g a diagram (Appendix HL) denoted here fancifully by $\frac{\partial g}{\partial X} = D_X(g)$ (the 'decomposition of X according to the map g') over the small category ΓY as follows:

$$D_X(g)(\sigma) = g^{-1}(\sigma) \subseteq X.$$

Thus $g^{-1}(\sigma)$ is the (strict) pullback in:

$$\begin{array}{ccc} g^{-1}(\sigma) & \longrightarrow & X \\ \downarrow & & \downarrow g \\ \Delta[n] & \xrightarrow{\sigma} & Y \end{array}$$

As we saw above in 1.F, X is weakly equivalent to the homotopy colimit over ΓY of $D_X(g)$. Symbolically $X \simeq \oint_{\Gamma Y} D_X(g)$. The map g itself can be recovered from $D_X(g)$ by taking the unpointed homotopy colimit of its map into a diagram $(*)$ of points in the shape of ΓY.

But our point here is that $D_X(g)$ is the diagram of actual fibres or point-inverses of g. (Working with simplicial complexes one can actually take the point inverses of all the barycentres of linear simplexes in the base.) The diagram $D_X(g)$ is *unpointed*; a choice of base point in each $g^{-1}(\sigma)$ amounts to a choice of a homotopy section of g, a section that may not exist.

We now show that one can rearrange the spaces $g^{-1}(\sigma) \equiv D_X(g)(\sigma)$ over a contractible small category denoted by λY (it depends only on Y) in such a way that the (unpointed) homotopy colimit will be homotopy equivalent to $\mathrm{Fib}(X \to Y)$.

B.1 PROPOSITION: *For every map $g : X \to Y$ there exists a diagram $\lambda(g) : \lambda Y \to S$ of unpointed spaces such that:*

(a) For each $i \in \lambda Y$ the space $\lambda(g)(i)$ that appears in the diagram is equal to $g^{-1}(\sigma)$ for some $\sigma : \Delta[n] \to Y$.

(b) The small category λY is contractible.

(c) The unpointed homotopy colimit of $\lambda(g)$ is weakly equivalent to $\mathrm{Fib}(g)$, namely:

$$\oint_{\lambda Y} \lambda(g) \simeq \mathrm{Fib}(X \xrightarrow{g} Y).$$

Proof: Notice that g is not necessarily a fibre map so the homotopy type of $g^{-1}(\sigma)$ varies as with σ. We first give a few examples and draw important consequences, and then give the simple proof below. Compare section E.4 below.

B.1.1 EXAMPLE: Here is an illuminating example of how the homotopy fibre of a map g arises as a homotopy direct limit of a diagram consisting of the actual fibres, i.e. inverse images of g. Notice that in this example there is no difference whether one takes a pointed or unpointed homotopy colimit since the small category we will consider is linear and contractible. Consider any self-map $w : W \to W$ of an arbitrary space W. Form the mapping torus $T(w) \equiv W \times I/(x,0) \sim (w(x),1)$. The map $W \to *$ induces a map $g : T(w) \to S^1$ on mapping toruses since S^1 is the mapping torus of the identity $* \to *$. Consider the strict pullback diagram, which is also a homotopy pullback since the universal covering map over the circle is a fibre map:

$$\begin{array}{ccc} \mathrm{Fib}(g) & \longrightarrow & \mathbb{R} \\ \downarrow & & \downarrow \\ T(w) & \longrightarrow & S^1 \end{array}$$

This diagram gives the usual natural construction of the homotopy fibre of g, namely $\mathrm{Fib}(g)$ as a pullback from $\mathbb{R} \to S^1$. But a direct examination of the strict pullback $\mathrm{Fib}(g)$ in this square reveals that $\mathrm{Fib}(g)$ is in fact isomorphic (homeomorphic) to the infinite mapping telescope $\mathrm{Tel}(w)$ of the original map $w : W \to W$. This telescope is clearly equivalent to the homotopy colimit of the tower $(W \xrightarrow{w} W \xrightarrow{w} W \to \cdots)$. Now notice that W is equal to $g^{-1}(y)$ for any $y \in S^1$. So the homotopy fibre is presented here as an infinite telescope in terms of the strict inverse images of our map g.

Notice that, although the infinite telescope is not given as a pointed diagram or a pointed homotopy colimit, one can choose a consistent base point system, and pointed and unpointed homotopy colimits are equivalent here.

We now prove (B.1).

Proof of B.1: We consider the construction of $\mathrm{Fib}(g)$ as a strict pullback in:

$$\begin{array}{ccc} \mathrm{Fib}(g) & \xrightarrow{\tilde{g}} & \Lambda Y \\ \downarrow & & \downarrow \ell \\ X & \xrightarrow{g} & Y \end{array}$$

The space ΛY is the simplicial path space over the pointed space Y [May-1]. The space ΛY is a contractible space and we take the category λY to be $\Gamma(\Lambda Y)$. We define the diagram $\lambda(g)$ to be the functor $\lambda(g)\colon \lambda Y \to \mathcal{S}$ with $\lambda(g)(\varphi) = \tilde{g}^{-1}(\varphi)$. In other words, $\lambda(g)$ is the diagram of inverse images of the map $\tilde{g}$. Since $\Lambda Y \simeq *$, (B.1)(b) is satisfied. We now consider the map $\tilde{g}$; by the decomposition of the domain of a map as a homotopy colimit of the inverse images as discussed above, the domain of $\tilde{g}$ is given as: $\mathrm{Fib}(g) = \operatorname{hocolim}_{\sigma\in\Gamma\Lambda Y}\{g^{-1}(\sigma)\}$ so $\lambda(g)$ satisfies also (B.1)(c). To prove (B.1)(a) notice that, since the diagram defining $\mathrm{Fib}(g)$ is a strict pullback for any simplex $\tilde{\sigma} \in \Lambda Y$, we have an isomorphism:

$$(\tilde{g})^{-1}(\tilde{\sigma}) \simeq g^{-1}(\ell(\tilde{\sigma})).$$

In fact g and $\tilde{g}$ have *the same* point inverses over corresponding simplices. This completes the proof of (B.1). ∎

FROM UNPOINTED TO POINTED HOMOTOPY COLIMITS: Presenting the homotopy fibre as a homotopy direct limit construction <u>over a contractible category</u> out of the actual inverse images of simplices (or points) in the range, over a contractible small category, was our first step. The second step is a theorem which guarantees that, if space X is homotopy equivalent to $\oint_I \underset{\sim}{Y}$ with $I \simeq *$, and $Y(i) \neq \emptyset$, then X can also be obtained up to homotopy by repeated construction of a *pointed* homotopy colimit over a sequence of small categories J_α of pointed diagram $\underset{\sim}{Y}_{*,\alpha}$, which consists of the same spaces $\underset{\sim}{Y}(i)$ or spaces built from them by pointed hocolims:

B.2 THEOREM: *Let $\underset{\sim}{Y}\colon I \to \mathcal{S}$ be a diagram of non-empty space $\underset{\sim}{Y}(i) \neq \emptyset$ with $I \sim *$. There exists a choice of base points $* \in \underset{\sim}{Y}(i)$ with respect to which*

$$\operatorname*{hocolim}_I \underset{\sim}{Y} \in \mathcal{C}^{\cdot}\big(\bigvee_i \underset{\sim}{Y}(i)\big).$$

An outline of a proof is given after the proof of (B.3) below. The proposition says, loosely speaking, that any unpointed homotopy colimit over a contractible category can be obtained as a repeated pointed homotopy colimit using the same spaces with a proper choice of base points.

An immediate consequence of (B.1) and (B.2) is

B.3 COROLLARY: *Let $g\colon X \to Y$ be any surjective map of pointed simplicial sets (or complexes). If $\mathcal{U}$ is any closed class such that, for each $\sigma \in Y$, one has $g^{-1}(\sigma) \in \mathcal{U}$*

with respect to some choice of base point $* \in g^{-1}(\sigma)$, *then the homotopy fibre* $\mathrm{Fib}(g)$ *is also in the class* $\mathcal{U}$.

SLOGAN: Thus: 'The homotopy fibre $\mathrm{Fib}(g)$ can be built as a pointed homotopy colimit out of the 'real fibres'— the point inverses $\{g^{-1}(y)\}$.' In particular, as a special case we get the following result of Quillen [Q-2]:

B.3.1 COROLLARY: *If* $g\colon X \to Y$ *is a simplicial map with contractible point inverse* $g^{-1}(y) \simeq *$, *then* g *is a weak equivalence, namely its homotopy fibre is also contractible. Moreover if the point inverses are all* n*-connected spaces then so is the homotopy fibre.*

Non-example: The exponential map for topological spaces $exp : [0,1) \longrightarrow S^1$ has single points as all its point inverses but it is not an equivalence. As it stands it is not a simplicial map though.

Similarly one has:

B.3.2 COROLLARY: *If* $g\colon X \to Y$ *is a simplicial map with* $g^{-1}(y)$ *and an* E_**-acyclic space* E_* *is any homology theory, then the homotopy fibre is also* E_**-acyclic.*

Proof of B.3: Apply (B.1) to the map $g\colon X \to Y$. We get that $\mathrm{Fib}(g)$ is a homotopy colimit as in (B.1)(c). Now apply (B.2) to (B.1)(a), by using surjectivity which gives us a diagram of non-empty spaces $g^{-1}(\sigma)$, where $\sigma \in \Gamma Y$. By (B.2) we get $\mathrm{Fib}(g)$ as a pointed homotopy colimit, as needed.

Proof of B.3.1 and B.3.2: Since if $Y \simeq *$ every space in $C^{\cdot}(Y)$ is contractible, we get from (B.3) that $\mathrm{Fib}(g) \simeq *$. Similarly, if $\tilde{E}_*(Y) \simeq 0$ then $\tilde{E}_*(Y') \simeq 0$ for any $Y' \in C^{\cdot}(Y)$. So again we get (B.3.2) from (B.3)

Outline of proof of (B.2): [Am][Ch-2] We must show how to choose pointed versions of $\underset{\sim}{Y}(i)$ from which we construct $\operatorname{hocolim}_I \underset{\sim}{Y}$ by repeatedly taking pointed homotopy colimits. We give an outline of a proof based on the work of Amit [Am]. (An independent proof was given in [Ch-2] that actually has a stronger version and consequences.) Amit reduces everything to taking the colimit over finite collapsible categories that are the underlying categories of simplicial complexes.

The proof rests on two examples of unpointed direct limits over contractible small categories that can be easily made pointed.

B.4 LEMMA: *Theorem B.2 is true if* I *is either the category associated to a partially ordered set satisfying (with* $x \to y$ *being* $x < y$*)* $\forall x \forall y \exists z (x < z) \& (y < z)$ *(filtering indexing category) or the 'elementary pushout' category given by the diagram* $* \leftarrow * \rightarrow *$.

Proof of Lemma B.4: If $X_0 \leftarrow X_1 \rightarrow X_2$ is a diagram of non-empty spaces, we can always choose a base point consistently by starting with one in X_1. If $\underset{\sim}{X}\colon I \to S$ and

i is any fixed object in the filtering category I as above, any direct limit of $\underset{\sim}{X}\colon \to S$ is isomorphic to the direct limit of $\underset{\sim}{X}^{\geq i}\colon I^{\geq i} \to S$, where $I^{\geq i}$ is the subcategory of all elements greater than or equal to i. But in $X^{\geq i}$ we again can choose a base point by first taking an arbitrary point in $X(i)$ which we assume to be non-empty. Since the homotopy direct limit over I can always be written as a strict direct limit over the associated free diagram (see Appendix HL), we are done.

Next, given any contractible category $I \simeq *$ Amit has shown how to write any direct limit over I as a series of direct limits over categories which are special, i.e. are of the forms in Lemma B.4 above. This he does in four steps:

Step 1: We can always reduce the situation to diagrams over the category of simplices of a contractible simplicial complex.
Step 2: We can assume that our category is the category of a finite contractible simplicial complex.
Step 3: B.2 is true for the category of simplices of a collapsible simplicial complex [Co].
Step 4: We can assume that our category is the category of a finite *collapsible* simplicial complex.

B.5 Outline of the proofs of 1–4

Step 1: This rests on the following two claims:

1.a Claim: If $\underset{\sim}{X}\colon I \to S$ is a diagram over a small category I, then the $\underset{\sim}{X}$ induces as in [D-K-3, sec. 5] a diagram $\overline{\mathrm{sd}}\underset{\sim}{X}$, over the opposite of the *subdivision* of I denoted by $\overline{\mathrm{sd}}I$. The subdivision is a certain quotient category of finite chains of composeable arrows in I. See [D-K-3]. The colimit of $\overline{\mathrm{sd}}X$ over $\overline{\mathrm{sd}}I$ is naturally isomorphic to the colimit of $\underset{\sim}{X}$ over I. The following interesting combinatorial property of subdivision is crucial for the present approach:

1.b Claim: If I is any small category then the <u>third subdivision</u> of I, namely $(\overline{\mathrm{sd}})^3 I$, is the category associated with a simplicial complex.

Both claims are proved by a straightforward detailed computation.

Step 2: Any contractible simplicial complex is the direct limit of the partially ordered collection of all finite contractible simplicial subcomplexes. Therefore any direct limit over a given infinite complex can be written as a composition of direct limits over finite complexes followed by a direct limit over the partially ordered indexing category of all finite subcomplexes. Now we can use Lemma B.4 above, since this partially ordered set is 'filtered' as in (B.4).

Step 3: Again by direct computation one shows that, if a simplicial complex K collapses to L, then any diagram over K induces one over L and vice versa. Moreover, these diagrams will have naturally isomorphic direct limits. This is done inductively by collapsing one simplex at a time.

Given a diagram $\underset{\sim}{X}\colon K \to \mathcal{S}$, where K is a category of simplices of a complex and $\ell\colon K \searrow K_1$ is an elementary collapse, then with each diagram $\underset{\sim}{X}\colon K_1 \to \mathcal{S}$ over K_1 one can associate a diagram $\ell^*(\underset{\sim}{X})$ over K whose spaces over the extra (two) simplices are obtained from $\underset{\sim}{X}(i)$'s via an elementary pushout over the category $\cdot \leftarrow \cdot \rightarrow \cdot$. Moreover, there is a natural isomorphism between the two colimits:

Step 4: This follows from Step 3 when we recall that any contractible finite complex is the collapse of a finite collapsible complex.

This completes the proof of (B.2). ∎

C. The weak Cauchy-Schwartz inequality

Theorem B.3 above has a direct consequence that also gives us a generalization. Given a map of I-diagrams in $\mathcal{S}$ (or $\mathcal{S}_*$) $\underset{\sim}{g}\colon \underset{\sim}{X} \to \underset{\sim}{Y}$, one would like to estimate the homotopy fibre of the induced maps between the homotopy colimits, namely to estimate $\mathrm{Fib}(\operatorname{hocolim}_I \underset{\sim}{g})$. Since we are in an unpointed context we put ourselves in a connected context: We denote by $c\mathcal{S}$ or $c\mathcal{S}_*$ the sub-categories of connected spaces.

C.1 THEOREM: *Let $\underset{\sim}{g}\colon \underset{\sim}{X} \to \underset{\sim}{Y}$ be a map of I-diagrams in $c\mathcal{S}$ (corresponding $c\mathcal{S}_*$) and $\mathcal{W}$ a closed class. Assume that for each $i \in \mathrm{obj}I$ the homotopy fibre $\mathrm{Fib}(G(i))$ is connected and belongs to $\mathcal{W}$. Then the homotopy fibre of $\operatorname{hocolim}_I \underset{\sim}{g}$ is connected and belongs to $\mathcal{W}$:*

$$\mathrm{Fib}\left(\oint_I \underset{\sim}{X} \to \oint_I \underset{\sim}{Y}\right) \in \mathcal{W}.$$

(Correspondingly, the homotopy fibre of the pointed $\operatorname{hocolim}_I \mathrm{Fib}(\int_I \underset{\sim}{g})$ also belongs to $\mathcal{W}$.)

Proof: We may assume that for each $i \in \mathrm{obj}I$ the map $\underset{\sim}{g}(i)\colon \underset{\sim}{X}(i) \to \underset{\sim}{Y}(i)$ is a fibre map, otherwise we functorially turn all these maps into fibre maps. Since $\underset{\sim}{Y}(i)$ is connected $\underset{\sim}{g}(i)$ has a well-defined (up to homotopy) fibre and we assume that this fibre is connected and, with respect to any choice of base point, is a member of the closed class $\mathcal{W}$. Now we observe that for every simplex $\sigma \in \operatorname{hocolim}_I \underset{\sim}{Y}$, there is a simplex $\sigma_i \in \underset{\sim}{Y}(i)$ such that

$$(\underset{\sim}{g}(i))^{-1}(\sigma_i) \simeq g^{-1}(\sigma)$$

where $g = \operatorname{hocolim}_I \underset{\sim}{g}$ is the induced map on the homotopy colimits. This follows directly from the universal construction of hocolim in [B-K]. Therefore, by assumption for each $\sigma \in \operatorname{hocolim}_I \underset{\sim}{Y}$, we have $g^{-1}(\sigma) \in \mathcal{W}$. Hence by Theorem B.3, Fib(g) is connected and is a member of $\mathcal{W}$, as needed.

Once we have proved the unpointed version, the pointed version follows from it as a special case or can be proven by an analogous pointed version.

C.1.1 REMARK: Theorem B.3 is also a special case of (C.1). Consider any map $g: X \to Y$ in $\mathcal{S}$. The map g can be presented, up to homotopy, as a homotopy colimit of a map of diagrams as follows: we write X as the homotopy colimit of the ΓY-diagram $D_X(g)$, which is our notation for the diagram whose value at a simplex $\sigma \in \Gamma Y$ is $g^{-1}(\sigma)$. Now $\operatorname{hocolim}_{\Gamma Y}(pt) \simeq Y$ and $\operatorname{hocolim}_{\Gamma Y} D_X(g) = Y$.

Therefore, on taking the homotopy colimit, the map of diagrams $D_X(g) \to (pt)$ into the constant point-diagram induces the given map g. Further, the inverse image of a simplex $\sigma \in Y$ under g is precisely the homotopy fibre of $D_X(g)(\sigma) \to pt$. Therefore, if one applies (C.1) to the map $D_X(g) \to (pt)$, one gets (B.3).

D. Examples

D.1 THE FIBRE AND COFIBRE OF A MAP. For any map $g: E \to B$ to a connected space B with homotopy fibre F, we have $\Sigma F \ll B/E \simeq B \cup CE$. Stated otherwise, the cofibre of g is cellular with respect to the suspension of the fibre. This follows directly from WCS, as we saw earlier in B.2 (iii). In fact, one can write B/E as hocolim$_*$ of a ΓB-diagram consisting entirely of spaces homotopy equivalent to ΣF: Let $\underset{\sim}{F} = \underset{\sim}{F}(p)$ be the ΓB-diagram of the fibres of p; namely we take ΓB to be a small category whose classifying space is homotopy equivalent to B, e.g. ΓB can be taken to be the small category of (non-degenerate) simplices of B. Then $\underset{\sim}{F}$ is a functor $\underset{\sim}{F} : \Gamma B \longrightarrow \mathcal{S}$ with $\operatorname{hocolim}_{\Gamma B} \underset{\sim}{F} \simeq E$ and the natural map $\operatorname{hocolim}_{\Gamma B} \underset{\sim}{F} \to \operatorname{hocolim}_{\Gamma B} \{pt\}$ is the map $p : E \to B$ under the natural equivalences. The ΓB-diagram $\underset{\sim}{F}$ is not pointed. However, take the free suspension object-wise in $\underset{\sim}{F}$ to get

$$\Sigma \underset{\sim}{F} : \Gamma B \to \mathcal{S}.$$

The diagrams $\underset{\sim}{F}$ and $\Sigma \underset{\sim}{F}$ consist of spaces which are homotopy equivalent to F or ΣF and all the maps in the diagram are homotopy equivalences. Now $\Sigma \underset{\sim}{F}$ is pointed by (any) one of the suspension points. So consider $\Sigma \underset{\sim}{F} : \Gamma B \to \mathcal{S}_*$ as a pointed diagram.

CLAIM: *There is a natural homotopy equivalence*

$$\int_{\Gamma B} \Sigma \underset{\sim}{F} \xrightarrow{\simeq} B/E$$

that realizes the inequality $\Sigma F \ll B/E$.

To prove the claim we recall the relation between pointed and unpointed homotopy colimits [B-K]. For any pointed diagram $\underset{\sim}{X}$ we have a cofibration (2.D.3)

$$|I| \to \operatorname{hocolim}_I \underset{\sim}{X} \to \int_I \underset{\sim}{X} = \operatorname{hocolim}_{*I} \underset{\sim}{X}.$$

So to compute $\int_{\Gamma B} \Sigma \underset{\sim}{F}$ we first compute $\operatorname{hocolim}_{\Gamma B} \Sigma \underset{\sim}{F}$. Since

$$\Sigma F = \operatorname{hocolim}_{\cdot \leftarrow \cdot \rightarrow \cdot}(pt \leftarrow F \to pt),$$

thus $\operatorname{hocolim}_{\Gamma B} \Sigma \underset{\sim}{F}$ is, by commutation of hocolims:

$$\operatorname{hocolim}_{\cdot \leftarrow \cdot \rightarrow \cdot}(\operatorname{hocolim}_{\Gamma B} pt \leftarrow \operatorname{hocolim}_{\Gamma B} \underset{\sim}{F} \to \operatorname{hocolim}_{\Gamma B} pt)$$

$$\simeq D\mathrm{cyl}(B \leftarrow E \to B)$$

(because $\operatorname{hocolim}_{\Gamma B} pt \simeq B$). Therefore, since $\int_{\Gamma B} \Sigma \underset{\sim}{F} = \left(\operatorname{hocolim}_{\Gamma B} \Sigma \underset{\sim}{F}\right)/B$, we get

$$\int_{\Gamma B} \Sigma \underset{\sim}{F} \simeq D\mathrm{cyl}(B \leftarrow E \to B)/B = \operatorname{colim}(* \leftarrow E \to B)$$

$$\simeq B \cup_p CE$$

as needed.

Example: As a minor example of the use of the cellular inequality $\Sigma F \ll B/E$, here is a quick cellular proof of a well known theorem of Hopf: *For any connected space X the map $H_2X \to H_2K(\Pi, 1)$ is surjective.* Consider the fibration $\tilde{X} \to X \to K(\Pi, 1)$. We have an exact sequence on integral homology of a pair:

$H_2X \to H_2K(\Pi,1) \to H_2(K(\Pi,1),X)$. Now $H_2(K(\Pi,1),X) = H_2(K(\Pi,1)/X)$. First, notice that by the above inequality $\Sigma\tilde{X} \ll K(\Pi,1)/X$. Second, $\Sigma\tilde{X}$ is 2-connected as is any space built from it by a pointed homotopy colimit (2.D.2.5), so $K(\Pi,1)/X$ is 2-connected too, hence $H_2(K(\Pi,1),X) \simeq 0$. Therefore, by exactness, we get the desired surjection.

D.2 DURABLE OR STRONG HOMOLOGY ISOMORPHISMS. Given a map $p : X \to Y$ that induces an E_*-isomorphism with respect to some homology theory, it does not follow that its homotopy fibre is E_*-acyclic. If it is acyclic, then we have a special 'strong' homology isomorphism. Similarly, though the cofibre of p is acyclic its loop space $\Omega Y/X$ is not in general acyclic. If, however, the homotopy fibre is acyclic, then the loop on the cofibre is also an acyclic space with respect to E_*. This occurs, for example, for any suspension of an E_*-isomorphism. Thus the suspension of a homology isomorphism is always a 'strong' homology isomorphism.

(I) CLAIM: *If p above is an E_*-isomorphism, then the homotopy fibre of Σp is E_*-acyclic.*

Proof: We have a cofibration $Y/X \to \Sigma X \to \Sigma Y$ with Y/X acyclic. Therefore, by (A.10) above the fibre of Σp is Y/X-cellular, and thus also acyclic (2.D.2.5).

(II) CLAIM: *If $X \to B$ is an E_*-isomorphism with an E_*-acyclic homotopy fibre, then the loop space of the cofibre $\Omega(B/X)$ is E_*-acyclic.*

Proof: Let $\mathbf{P}_{E_*} = \mathbf{P}_A$ be the functor $(\ )^+_{E_*}$, namely the functor that kills precisely all E_*-acyclic spaces. We need to show $\mathbf{P}_{E_*}\Omega(B/X) \simeq *$.

$$\begin{aligned}\mathbf{P}_A\Omega(B/X) &\simeq \Omega\mathbf{P}_{\Sigma A}(B/X)\\ &\simeq \Omega\mathbf{P}_{\Sigma A}\int_{\Gamma B} \Sigma \underset{\sim}{F},\end{aligned}$$

where F is the homotopy fibre of $X \to B$. We continue (1.D):

$$\simeq \Omega\mathbf{P}_{\Sigma A}\int_{\Gamma B} \mathbf{P}_{\Sigma A}\Sigma\underset{\sim}{F}.$$

Now since $\mathbf{P}_A\underset{\sim}{F} \simeq\sim *$ by assumption on F, we get by (1.A.8)(e.10) $\mathbf{P}_{\Sigma A}\Sigma F \simeq *$. Thus

$$\int_B \mathbf{P}_{\Sigma A}\Sigma\underset{\sim}{F} \simeq * \quad \text{and} \quad \mathbf{P}_A\Omega(B/X) \simeq *$$

as needed.

D.3 POINTED BOREL CONSTRUCTION. We now give the classifying space $\overline{W}G$ of any topological group G, as a pointed homotopy colimit of the suspension of the group ΣG with respect to the natural action of G:

We saw in (3.B.3) above that for a connected space X we always have $\Sigma\Omega X \ll X$. In other words, there exists a process of repeatedly taking homotopy colimits that leads from ΣG to the classifying space $\overline{W}G$. An explicit construction of X as pointed hocolim out of $\Sigma\Omega X$ is given as follows. Consider the map $* \to X$ into a connected X. The homotopy fibre being ΩX and cofibre being X itself, we get as in (D.2) above:

$$X = \int_{\Gamma X} \Sigma\Omega X.$$

We claim that one can replace the domain ΓX by the topological category ΩX. Pick a model for ΓX to be a topological (or simplicial) category with one object $\bullet$ and $\mathrm{map}(\bullet, \bullet) = \Omega X$ as a simplicial group. Then the classifying space of the (small) topological category ΩX is $|\Omega X| \simeq X$. So ΩX as a topological category is a good model for (ΓX): a diagram with $\operatorname{hocolim}_{\Gamma(\Omega X)}(pt) \simeq X$.

Therefore using half-smash:

$$X \simeq \int_{\Gamma X} \Sigma\Omega X \simeq \int_{\Omega X} \Sigma\Omega X \simeq E\Omega X \ltimes_{\Omega X} \Sigma\Omega X.$$

In other words, if $\Omega X = G$ is any topological group, then $X = \overline{W}G$ in the above considerations yields:

$$\overline{W}G = EG \ltimes_G \Sigma G.$$

In particular, we get (see A.6 above):

D.4 PROPOSITION: *For any group G the following implication holds:*

$$\mathrm{map}_*(\Sigma G, Y) \simeq * \Rightarrow \mathrm{map}(\overline{W}G, Y) \simeq *, \quad \textit{in other words } \Sigma G < \overline{W}G.$$

REMARK: The last implication follows also from (3.A), since $\mathbf{P}_{\Sigma G}\overline{W}G \simeq *$.

D.5 Corollary: *In any cofibration $A \to X \xrightarrow{j} X/A$ the suspension of the fibre F of j is null-equivalent to ΣA. Thus for all W*

$$\mathrm{map}_*(\Sigma F, W) \simeq * \Leftrightarrow \mathrm{map}_*(\Sigma A, W) \simeq * \quad \text{or} \quad \Sigma A < \Sigma F < \Sigma A.$$

Proof: Since $\mathrm{cof}(j) = \Sigma A$, we have, by (D.1) above, $\Sigma F < \Sigma A$. But we saw (A.10) that $A \ll F$, so by (A.6) $A < F$ and thus by (1.A.8)(e.10) $\Sigma A < \Sigma F$.

D.6 Example: For any fibration $F \to E \to B$ over a connected B and any connected X, the Theorem C.1 asserts

$$F \ll \mathrm{Fib}(E \vee X \to B \vee X).$$

To see why, we simply consider the diagram of fibrations:

$$\begin{array}{ccccc}
F & \longleftarrow & pt & \longrightarrow & pt \\
\downarrow & & \downarrow & & \downarrow \\
E & \longleftarrow & pt & \longrightarrow & X \\
\downarrow & & \downarrow & & \| \\
B & \longleftarrow & pt & \longrightarrow & X
\end{array}$$

It follows that the fibres of $E \to B$ and $E \vee X \to B \vee X$ are cellular-equivalent to each other.

Similarly

$$\mathrm{Fib}(X \to Y) \ll \mathrm{Fib}(X/V \to Y/V)$$

for any $V \to X \to Y$.

D.7 Cellular version of Blakers–Massey theorem: On the level of homotopy groups this classical theorem asserts that in a certain range, depending on the connectivity, the natural map $\pi_i(X, A) \longrightarrow \pi_i(X/A)$ is an isomorphism: If $\mathrm{conn} A = s-1$ and $\mathrm{conn}(X, A) = n-1$ then the latter map is an $n+s-1$-isomorphism (see [G, 16.30]). Reformulating in terms of homotopy fibre the result can be viewed as a statement about the connectivity of the map $F \longrightarrow A$ from the fibre F of $X \longrightarrow X/A$ to A itself. A recent result of Chacholsky gives a cellular version of Blakers–Massey theorem that is much stronger than the original:

PROPOSITION D.7.1 [CH-4]: *Let $A \longrightarrow X$ be a cofibration and F be the homotopy fibre of the map $X \longrightarrow X/A$. If F is connected then the following inequality holds:*

$$A * \mathrm{Fib}(A \longrightarrow X) \ll \Sigma(F/A)$$

We will not reproduce the proof that uses WCS above. ∎

Since cellular inequality preserves connectivity the classical Blakers–Massey theorem follows without difficulty. However this inequality holds without higher connectivity assumptions and estimates, in general, from below, the 'difference' between the fibre and the cobase of $X \longrightarrow X/A$ in term of A and $Fib(A \longrightarrow X)$. In [Ch-4] a more general theorem is proven comparing homotopy pushouts and homotopy pullback is general squares.

E. Average or weak colimit of a diagram

In Proposition B.1 above the homotopy fibre of an arbitrary simplicial map is constructed out of the strict inverse images of simplices by taking a certain homotopy colimit over a contractible category associated to the range of the map. In this section we want to put this construction in a wider perspective: Given any diagram of objects in any small category one may want to turn this diagram, in a universal way, into a diagram of isomorphisms or equivalences. Thus we might want to turn a diagram of groups into a diagram of group isomorphisms in a universal way, or a diagram of sets into a diagram of set isomorphisms. This reminds one of the process of taking a direct limit (i.e. colimit). Forming a colimit can be viewed as a process of turning a diagram $\mathcal{X}: I \longrightarrow \mathcal{E}$ given over a category $\mathcal{E}$ into a diagram of identities. If we relax the demand and ask for isomorphisms rather than identities we get a construction that might be called a weak colimit or average.

E.1 EXAMPLES

1. Consider the small category $D = *\rightrightarrows*$ consisting of two objects and two non-identity arrows between them. So the classifying space of D is the circle. and let $\mathcal{X} = \{\emptyset \rightrightarrows x_0\}$ be a diagram over D consisting of the empty set mapping twice to a single point x_0. The colimit of $\mathcal{X}$ is a single point, but the weak colimit or average av$\mathcal{X}$ of $\mathcal{X}$ consists of a diagram of two infinite sets mapping via both arrows by isomorphisms. Notice that no diagram consisting of two isomorphisms between two finite sets can be universal among all such i.e. in the present case, has a map to all such. It is not accidental that we get here the loop space of the circle as equivalent to the average.

2. Consider the category $\mathcal{T}$ with one object and an infinite number of non-identity self-maps $\{g^n = g \circ g \cdots \circ g : n > 0\}$, namely mor$I$ = $\mathbb{N}$, the natural

numbers. Then if we take N itself as a set on which $\mathcal{T}$ acts by shifting n to $n+1$ we see that g acts by a non-isomorphism. Turning this action in a canonical way into isomorphisms we get that the weak colimit is the diagram consisting of the integers Z with g acting as a shift map which is now a non-identity isomorphism. In short $\mathrm{av}\mathcal{T} = \mathbf{Z}$.

E.2 DEFINITION: *Given a diagram* $\mathcal{X}\colon I \longrightarrow \mathbf{E}$ *the diagram* $\mathrm{av}\mathcal{X}\colon I \longrightarrow \mathbf{E}$ *will be called a weak limit or average of* $\mathcal{X}$ *if all the maps in* $\mathrm{av}\mathcal{X}$ *are isomorphisms and it comes with coaugmentation* $\mathcal{X} \longrightarrow \mathrm{av}\mathcal{X}$ *that is a universal map of* $\mathcal{X}$ *into diagrams of isomorphisms over* I.

Remark: Thus av is the left adjoint to the inclusion functor from diagrams of isomorphisms over I to all diagrams over I. Such an adjoint exists whenever the category $\mathbf{E}$ is co-closed, i.e. has arbitrary direct limits (=colimits).

AVERAGE AS A LEFT KAN EXTENSION. It might be illuminating to present the weak colimit as a left Kan extension [Mac]. Consider the (non-commutative) diagram, where the diagonal is the Kan extension of the vertical arrow.

$$
\begin{array}{ccc}
I & \longrightarrow & I[I^{-1}] \\
{\scriptstyle \mathcal{X}}\Big\downarrow & {\Large\swarrow}{\scriptstyle I^{-1}\mathcal{X}} & \\
\mathbf{E} & &
\end{array}
$$

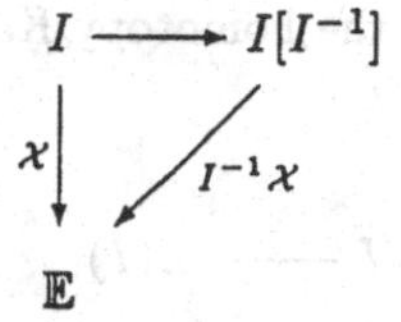

By $I[I^{-1}]$ we denote as usual the localization of I with respect to all its arrows, namely, the category obtained by formally adding inverses to all arrows in I. Assuming $\mathbf{E}$ is co-closed we always have a canonical extension of the functor $\mathcal{X}\colon I \longrightarrow \mathbf{E}$ to the category $I[I^{-1}]$. Since all morphisms in the latter are isomorphisms the diagram obtained over the localization of I consists of isomorphisms in $\mathbf{E}$. Now one can construct the average of the given $\mathcal{X}$ by precomposing into I.

E.3 DEFINITION:

$$
\mathrm{av}\mathcal{X} = \text{ the compositon } \quad I \xrightarrow{\ \mathcal{X}\ } I[I^{-1}] \xrightarrow{\ I^{-1}\mathcal{X}\ } \mathbf{E}.
$$

It now follows from the general properties of left Kan extensions that $\mathrm{av}\mathcal{X}$ is in fact universal in the above sense. From the point of view of the present section and (B.1)(b) above the following is a crucial property of the weak colimit construction.

PROPOSITION: *For each $i \in I$ the object in $\mathbb{E}$ given by $\mathrm{av}\mathcal{X}(i)$ is obtained as a colimit of a contractible diagram of objects in $\mathbb{E}$ of the form $\mathcal{X}(\alpha)$ for $\alpha \in I$.*

Proof: See [Mac]. The value of the Kan extension is obtained by taking colimit over certain *undercategory* which are always contractible, having an initial object.

E.4 SIMPLICIAL LOCALIZATION, HOMOTOPY AVERAGING AND FIBRATIONS. We now extend the notion of weak colimit to its homotopy version. Given a diagram of spaces we would like to turn it in a canonical way into a diagram of weak homotopy equivalences. Thus if the diagram comes from a simplicial map as in (B.1) above, once we turn it into a diagram of equivalences the typical homotopy type appearing in the homotopy average diagram will be that of the homotopy fibre of the map we started with. Again, this typical value of the homotopy average will be obtained by taking a homotopy colimit over a contractible category. Thus we see that turning a map into a fibre map is the universal construction that turns this map into a map with homotopy equivalent inverse images (fibres). The actual construction is analogous to the above except that one uses the Dwyer-Kan simplicial localization [D-K-1] $L(I)$ of the category I rather than its localization $I[I^{-1}]$.

Explicitly, we define the homotopy averaging as the composition $I \longrightarrow L(I) \longrightarrow S$ in the triangle below that gives the homotopy Kan extension $L(\mathcal{X})$ of the given diagram of spaces $\mathcal{X}$:

$$\begin{array}{ccc} I & \longrightarrow & L(I) \\ {\scriptstyle \mathcal{X}}\downarrow & \swarrow{\scriptstyle L(\mathcal{X})} & \\ \mathcal{S} & & \end{array}$$

We can therefore conclude that the homotopy fibre of a simplicial map $X \longrightarrow Y$ to a connected space Y is the homotopy average of the diagram of the actual fibres defined over the small category ΓY associated to Y.

F. A list of questions

Below we list a rather random list of open problems relating to cellular constructions. Problem 2–3 seem hard and interesting. At present the only known case is where A is a circle or a higher sphere. It is curious that Problem 7 does not seem to be easy.

1. Develop arithmetic square techniques for P_A, L_f.

2. Give a recognition principle for a space Y to be equivalent to a function complex $\mathrm{map}_*(A, X)$ for a given A. For $A = S^1$ this leads e.g. to the Segal loop machine from Chapter 3.

In particular, for a given X and A, when is there a space Y with $P_A(X^B) = Y^B$?

3. Is it possible to recover $\mathbf{CW}_A(X)$ from the function complex $X^A = \text{map}_*(A, X)$ and A itself in a natural way. Of course here the function complex must be equipped with some extra structure such as the action of A^A on it.

4. For a non-connected homology theory give a construction of all acyclics or a large subclass thereof.

5. Prove: if X is a p-complete polyGEM then $[B\mathbb{Z}/p\mathbb{Z}, X]$ is not trivial.

6. Discuss the effect of $\mathbf{P}_A$, $\mathbf{L}_f$ on the fundamental group. In particular the effect of K-theory localization should be the same as that of homology with the corresponding coefficients. In general the effect on the fundamental group of any homology theory localization might be the same as that of the corresponding connective theory localization.

7. Prove or disprove: For any map f and pointed A, If X is nilpotent then so are $\mathbf{L}_f X$ and $\mathbf{CW}_A X$; the same question for X a PolyGEM or X a 1-connected space.

8. For a given generalized homology discuss the associated plus construction that kills all the acyclics: $\mathbf{P}_{acyclics}$.

9. Give necessary and sufficient conditions for a space to be cellular-equivalent to the n-sphere.

10. Prove or disprove: if X, Y are A-cellular then $X^A \vee Y^A \ll (X \vee Y)^A$.

11. Can one show that $B\mathbb{Z}/p\mathbb{Z} \ll BG_p$ for any connected compact Lie G? Dwyer [Dw-1] shows by a Jackowsky–McClure induction that $\mathbf{P}_{B\mathbb{Z}/p}BG \simeq BG[1/p]$, McGibbon has considered a similar question recently [McG, 2.1].

12. Let Z_s be the Bousfield–Kan tot_s-functor. Is $X \ll Z_s X$ or at least $X < Z_s X$ for $s < \infty$?

13. Give conditions on a (coaugmented) homotopy functor that will ensure that it can be turned into a simplicial one by a universal arrow.

14. Is it true that $X \ll Y$ implies $SP^k X \ll SP^k Y$? True for $k = \infty$.

□□□

Appendices

Appendix HL: Homotopy colimits and fibrations

We give here a brief review of homotopy colimits and, more importantly, we list some of their crucial properties relating them to fibrations. These properties came to light after the appearance of [B-K]. The most important additional properties relate to the interaction between fibration and homotopy colimit and were first stated clearly by V. Puppe [Pu] in a paper dealing with Segal's characterization of loop spaces. In fact, much of what we need follows formally from the basic results of V. Puppe by induction on skeletons. Another issue we survey is the definition of homotopy colimits via free resolutions.

The basic *definition of homotopy colimits* is given in [B-K] and [S], where the property of homotopy invariance and their associated mapping property is given; here however we mostly use a different approach. This different approach from a 'homotopical algebra' point of view is given in [DF-1] where an 'invariant' definition is given. We briefly recall that second (equivalent) definition via free resolution below. Compare also the brief review in the first three sections of [Dw-1]. Recently, two extensive expositions combining and relating these two approaches appeared: a shorter one by Chacholsky and a more detailed one by Hirschhorn.

Basic property, examples

Homotopy colimits in general are functors that assign a space to a strictly commutative diagram of spaces. One starts with an arbitrary diagram, namely a functor $\underset{\sim}{X}\colon I \to$ (Spaces) where I is a small category that might be enriched over simplicial sets or topological spaces, namely in a simplicial category the morphism sets are equipped with the structure of a simplicial set and composition respects this structure. To such a diagram the functor hocolim (otherwise called homotopy direct limit and denoted $\underrightarrow{\text{holim}}_I$) assigns a space $\operatorname{hocolim}_I \underset{\sim}{X} \in$(Spaces). A basic property of this assignment is that it respects 'local weak equivalence of diagrams'.

PROPOSITION: *Let $f\colon \underset{\sim}{X} \to \underset{\sim}{Y}$ be a map of I-diagram of spaces. Thus f is a natural transformation between two functors $\underset{\sim}{X}$ and $\underset{\sim}{Y}$. Assume that for each $i \in I$ the map $\underset{\sim}{X}_i \to \underset{\sim}{Y}_i$ is a weak equivalence. Then f induces a weak equivalence of spaces*

$$\operatorname*{hocolim}_I f\colon \operatorname*{hocolim}_I \underset{\sim}{X} \longrightarrow \operatorname*{hocolim}_I \underset{\sim}{Y}.$$

In some sense the functor hocolim is the 'closest one' to the actual direct limit or colimit that has this 'weak homotopy invariant' property.

One can consider the functor hocolim in various categories $\{Spaces\}$ of spaces that have an appropriate notion of weak equivalence or 'homotopy of maps' (for example, in chain complexes or in simplicial (abelian or not) groups). We will need to consider it only in two cases: pointed spaces $\mathcal{S}_*$ and unpointed spaces $\mathcal{S}$. It turns out that their values on these categories are different, as can be seen in the following examples:

Examples:

(1) The most elementary examples of homotopy colimit are disjoint unions of (unpointed) spaces or wedges of pointed spaces. These are homotopy colimits over discrete categories, i.e. categories with only identity morphisms. A diagram over such a category is just a family of spaces. A pointed diagram is a family of pointed spaces. In this case the unpointed homotopy colimit is equal to the colimit itself. Already here the pointed and unpointed homotopy colimits are different. As always they are related by the cofibration $BI \to \operatorname*{hocolim}_I \underset{\sim}{X} \to \operatorname*{hocolim}_{*I} \underset{\sim}{X}$ so that in the pointed case one gets the wedge $\bigvee X_i$ as the pointed homotopy colimit over categories with no non-identity maps..

(2) If the category $I = G$ is a (discrete or continuous) group, then a diagram over G is a space with a group action and a pointed diagram is a space with a specified fixed point (fixed under the action of G). The homotopy colimit in the unpointed case is weakly equivalent to the Borel construction $EG \times_G X$, and this is true whether X has fixed points or not. In fact X may well be a single point $*$ and then $\operatorname{hocolim}_G(*) \simeq BG$, the classifying space of G. If G here is a simplicial or a topological category, we get BG as a topological or simplicial space by the definition of homotopy colimit. If X is a pointed G-space, then $\operatorname{hocolim}_* X$, the pointed homotopy colimit, is the pointed Borel construction given as $EG \ltimes_G X = (EG \ltimes X)/G$ where by $(-/G)$ we mean here the quotient space with respect to the G-action.

In contrast to the above the pointed Borel construction on a single point is contractible: $EG \ltimes_G \{*\} \simeq \{*\}$. (See Chapter 2 for a discussion of the half-smash product $(-) \ltimes (-)$.)

A theorem of V. Puppe

There are very useful relations between the concepts of 'fibre map', 'homotopy fibre' and that of homotopy colimits. These relations are useful since *sometimes* they allow one to commute the operation of taking the homotopy fibre of a map with that of taking homotopy colimits of a system of maps. In general one does not expect a 'left adjoint', such as a homotopy colimit, to commute with a 'right adjoint', such as taking a homotopy fibre of a map. In fact, in general, these operations do not commute.

Example: The homotopy colimit of the two-arrow pushout diagram $* \longleftarrow X \longrightarrow *$ is, of course, the suspension ΣX of X. Now given a map $g\colon X \to Y$ with Y connected, taking its homotopy fibre does not commute with suspension $\Sigma(\mathrm{Fib} g) \neq \mathrm{Fib}(\Sigma g)$. But notice that taking homotopy groups commutes with linear homotopy colimits, namely in a tower of cofibrations of pointed spaces $\underset{\sim}{X}$:

$$\underset{\sim}{X} = (X_0 \hookrightarrow X_1 \hookrightarrow X_2 \hookrightarrow \cdots)$$

one has

PROPOSITION: *For a linear tower as above there is a natural isomorphism*

$$\operatorname{colim} \pi_i \underset{\sim}{X} \to \pi_i \operatorname*{hocolim}_I \underset{\sim}{X}.$$

Proof: This follows from the compactness of S^i and the unit interval. ∎

Now when we translate this to properties of homotopy fibres using long exact sequences and the fact that linear direct limits preserve exactness, we get

PROPOSITION: *Let*

$$\begin{array}{ccccccccc} E_0 & \to & E_1 & \to & E_2 & \cdots & E_n & \cdots \to & E_\infty \\ p_0 \downarrow & & \downarrow & & \downarrow & & p_n \downarrow & & p_\infty \downarrow \\ B_0 & \to & B_1 & \to & B_2 & \cdots & B_n & \cdots \to & B_\infty \end{array}$$

be a tower of fibrations over connected pointed spaces B_i.

There is a weak equivalence relating homotopy fibres as follows

$$\operatorname*{hocolim}_I F_i \longrightarrow \mathrm{Fib}(\operatorname*{hocolim}_I p_i) \simeq \mathrm{Fib}(p_\infty\colon E_\infty \longrightarrow B_\infty).$$

More surprising perhaps is that under certain conditions homotopy pushout also commutes with taking homotopy fibre [Pu]:

THEOREM (V. PUPPE): *Let*

$$\begin{array}{ccccc} E_2 & \xleftarrow{\bar{f}_2} & E_0 & \xrightarrow{\bar{f}_1} & E_1 \\ \downarrow{\scriptstyle p_1} & & \downarrow{\scriptstyle p_0} & & \downarrow{\scriptstyle p_1} \\ B_2 & \xleftarrow{f_2} & B_0 & \xrightarrow{f_1} & B_1 \end{array}$$

be a commutative diagram and suppose that each square is a homotopy pullback, i.e. both $(f_1, \bar{f}_1)$ *and* $(f_2, \bar{f}_2)$ *induce a homotopy equivalence as the homotopy* $\mathrm{Fib}(p_0) \simeq \mathrm{Fib}(p_1) \simeq \mathrm{Fib}(p_2)$ *via the obvious maps. Then the maps of* $\mathrm{Fib}(p_i)$ *for* $i = 0, 1, 2$ *into the homotopy fibre of the maps of homotopy pushouts, namely*

$$\mathrm{Fib}(p_i) \xrightarrow{\simeq} \mathrm{Fib}(E_1 \tilde{\cup}_{E_0} E_2 \longrightarrow B_1 \tilde{\cup}_{B_0} B_2),$$

are all weak equivalences, where $\tilde{\cup}$ *denotes the homotopy pushout, i.e. double mapping cylinder.*

<u>General homotopy colimits of fibrations</u>

Since by induction one can build any homotopy colimit using direct towers, disjoint unions and pushouts, we can reach the following conclusion:

PROPOSITION [PU]: *Let* $\underset{\sim}{E} \to \underset{\sim}{B}$ *be any map of* I*-diagrams with* $\underset{\sim}{B}_i$ *connected for all* $i \in I$*. Assume that for each* $\varphi\colon i \to j$ *in mor* I *the square:*

$$\begin{array}{ccc} \underset{\sim}{E}_i & \longrightarrow & \underset{\sim}{E}_j \\ \downarrow & & \downarrow \\ \underset{\sim}{B}_i & \longrightarrow & \underset{\sim}{B}_j \end{array}$$

is a pullback square up to homotopy, so that the induced map on the homotopy fibres:

$$F_i = \mathrm{Fib}(\underset{\sim}{E}_i \longrightarrow B_i) \quad \xrightarrow{\simeq} \quad \mathrm{Fib}(\underset{\sim}{E}_j \longrightarrow \underset{\sim}{E}_j) = F_j$$

is a weak equivalence. Suppose in addition, for simplicity, that I *is connected (i.e.* BI *is a connected space). In this situation the homotopy fibre of the induced map on the homotopy colimits is equivalent to the common values of all the homotopy fibres, i.e.*

$$\mathrm{Fib}\Big(\underset{I}{\operatorname{hocolim}}\underset{\sim}{E} \longrightarrow \underset{I}{\operatorname{hocolim}}\underset{\sim}{B}\Big)$$

is weakly equivalent to the 'common value' F_i.

<u>Diagrams over a fixed base space B</u>

The above theorem of V. Puppe has a very useful corollary, where instead of considering a diagram of fibrations with 'fixed homotopy type as a homotopy fibre' one considers diagrams of fibration sequences over a fixed base space B. See the examples following the next proposition.

PROPOSITION: *Let $\underset{\sim}{E} \to B$ be any map of an I-diagram to a fixed connected space B. Thus all we assume is that all the squares strictly commute:*

$$\begin{array}{ccc} E_i & \longrightarrow & E_j \\ \downarrow & & \downarrow \\ B & \xrightarrow{=} & B \end{array}$$

Let $\underset{\sim}{F}$ be the diagram of the homotopy fibres of the maps $E_i \longrightarrow B_i$. In this situation the homotopy fibre of the map $\underset{I}{\operatorname{hocolim}}\underset{\sim}{E} \longrightarrow \underset{I}{\operatorname{hocolim}}\underset{\sim}{B} = B$ is equivalent to the homotopy colimit of the diagram of fibres: $\underset{I}{\operatorname{hocolim}}\underset{\sim}{F}$.

Proof: It is not hard to reduce this to the theorem of V. Puppe above by backing up the fibration to get a diagram of fibrations with a fixed homotopy type ΩB as the homotopy fibre. From the proposition above we get a fibration

$$\Omega B \longrightarrow \underset{I}{\operatorname{hocolim}}\underset{\sim}{F} \longrightarrow \underset{I}{\operatorname{hocolim}}\underset{\sim}{E}$$

that can be seen to be a principal one. Upon classifying this fibration sequence we get the desired fibration sequence over B. An easy direct proof can be obtained by decomposing each fibre map $E_i \longrightarrow B$ into a commutative diagram of trivial fibrations $F_i \longrightarrow (E_i)_\sigma \longrightarrow \sigma$ over the diagram of simplices $\{\sigma\}=\Gamma B$ as explained below. This gives a 'double diagram' of fibrations $F_{i,\sigma} \xrightarrow{\simeq} E_{i,\sigma} \longrightarrow \sigma$. The original diagram is obtained by 'integrating', i.e. taking homotopy colimit over the index σ that varies over the simplices of B. But since we are allowed to change the order of taking homotopy colimits we first take the homotopy colimit over the index $i \in I$.

Namely, we take first the hocolim over each simplex in B. This gives us a diagram of maps:

$$(*) \qquad \big(\operatorname*{hocolim}_I F_{i,\sigma} \xrightarrow{\simeq} \operatorname*{hocolim}_I E_{i,\sigma} \longrightarrow \sigma\big)_{\sigma\in\Gamma B}.$$

Notice the weak equivalence between fibre and total space that follows from the fact that for each individual i the map is an equivalence. Now as we vary σ over the diagram ΓB of simplices in B, for each fixed i a map $\sigma \to \sigma'$ induces an equivalence

$$(**) \qquad F_{i,\sigma} \xrightarrow{\simeq} F_{i,\sigma'}$$

because for each i the map $E_i \to B$ is a fibration. Therefore, by the basic property of homotopy colimit, upon taking homotopy colimit over $i \in I$ one still gets an equivalence $(*)$, as claimed. Now $(*)$, being a diagram of equivalences with respect to all maps in ΓB, it gives the desired fibration sequence over B upon taking hocolim with respect to $\sigma \in \Gamma B$.

EXAMPLES: Let I be the small diagram $\cdot \leftarrow \cdot \rightarrow \cdot$ and consider the diagram of maps:

$$\begin{array}{ccccc} X & \longleftarrow & (pt) & \longrightarrow & X \\ \downarrow & & \downarrow & & \downarrow \\ X & \xleftarrow{=} & X & \xrightarrow{=} & X \end{array}$$

Or in other words:

$$\begin{array}{ccccc} X & \longleftarrow & (pt) & \longrightarrow & X \\ & \searrow & \downarrow & \swarrow & \\ & & X & & \end{array}$$

Then the fibration of the hocolim is the fold map $X \vee X \to X$. In this case we get Ganea's theorem that the fibre is $\Sigma\Omega X$ since the diagram of fibres is $pt \longleftarrow \Omega X \longrightarrow pt$, whose hocolim is $\Sigma\Omega X$. Similarly, by considering the analogous diagram over $B \times B$ one gets that the homotopy fibre of $X \vee X \longrightarrow X \times X$ is the join $\Omega X * \Omega X$.

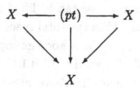

Another special case is the fibre of the map $E \cup CF \longrightarrow B$ for a fibration $F \longrightarrow E \longrightarrow B$. Here we use the diagram

$$\begin{array}{ccccc}
\Omega B & \longleftarrow & F \times \Omega B & \longrightarrow & F \\
\downarrow & & \downarrow & & \downarrow \\
* & \longleftarrow & F & \longrightarrow & E \\
\downarrow & & \downarrow * & & \downarrow \\
B & \overset{=}{\longleftarrow} & B & \overset{=}{\longrightarrow} & B
\end{array}$$

So the hocolim of the fibres is the join $F * \Omega B$. Notice that a special case of the fibration $\Sigma\Omega X \to X \vee X \to X$ is the wonderful fibration sequence:

$$S^2 \to \mathbb{C}P^\infty \vee \mathbb{C}P^\infty \to \mathbb{C}P^\infty$$

or

$$S^2 \to K(\mathbb{Z}, 2) \vee K(\mathbb{Z}, 2) \to K(\mathbb{Z}, 2).$$

Decomposing general maps into free diagrams

A particular case of V. Puppe's theorem, from which in fact the general case follows, is especially useful. This is the case when all the base spaces are contractible and thus all the spaces E_i are equivalent to each other and are just the fibres themselves. This case arises naturally when a fibration over, say, a simplicial complex is decomposed into a diagram of the simplices of the base spaces. Compare (1.F).

We want to think about fibration as a map $E \xrightarrow{p} B$ in which all the fibres $p^{-1}(b)$ are homotopy equivalent to each other; more correctly, a map in which each path $\varphi : I \to B, \varphi(0) \sim \varphi(1)$ in the base can be lifted to a homotopy equivalence of the fibres over these end points. Thus the total space of a fibre map is 'a family of equivalent fibres' glued together over B. A good example to consider is the mapping torus of a self-map $w : W \to W$ discussed in (9.B.1.1), where other examples of this procedure are discussed. Now to make the above more formal we work, for simplicity, with simplicial complexes, but everything can be done similarly with simplicial sets [Ch-3], [HH].

Consider an arbitrary simplicial map of simplicial complexes: $f : E \to B$. For the moment we do not assume that the map f is a fibre map. One can consider the small category defined by the simplices of the simplicial complex B. This category, denoted by ΓB, has one object σ for each simplex in B and one morphism $\sigma \to \tau$ for each face inclusion of simplices in B. Thus $\mathrm{Hom}(x, y)$ in ΓB has at most one element and usually it is empty. Now define a diagram $\underset{\sim}{E}$ of simplicial complexes indexed by ΓB as follows: $\underset{\sim}{E}_\sigma = f^{-1}(\sigma)$; this is a subcomplex of E. Thus

$\underset{\sim}{E}$ is simply the diagram of the inverse images of simplices in B. Notice that it is immediate from the definition of $\underset{\sim}{E}$ that its strict colimit over ΓB is equal to E itself: $\operatorname{colim}_{\Gamma B} \underset{\sim}{E} = E$.

The following is very useful (explanation below):

PROPOSITION: *In the situation above for any simplicial map $f : E \to B$:*

(1) *The diagram $\underset{\sim}{E}$ is a free ΓB-diagram.*

(2) *The homotopy colimit of $\underset{\sim}{E}$ is equivalent to the space E itself,* $\operatorname{hocolim}_I \underset{\sim}{E} \simeq E$, *when the homotopy colimit is taken over the indexing diagram ΓB.*

In order to explain this proposition we recall the treatment of homotopy colimit via free resolution of diagrams.

Recall: Free I-diagrams

By 'free diagram of spaces' we mean a 'cellular' diagram that 'can be built from free I-cells of the form $\Delta[n] \times F^d$ by disjoint unions and elementary pushouts'. Let us explain [DF-1]:

(i) An *orbit* is an I-diagram e whose strict direct limit is a point.

(ii) With each object $d \in I$ one associates an orbit denoted F^d and called the *free orbit at d*. By definition F^d: $I \to S$ is given by $F^d(d') = \operatorname{Hom}_I(d, d')$, where composition in I gives obvious maps $F_i^d \to F_j^d$ for each $(i \to j) \in \operatorname{mor} I$.

Example: Thus for the category $I = G$ associated to a group there is only one object and only one type of free orbit $F^\bullet = G = \operatorname{map}_G(\bullet, \bullet)$, namely the group itself with the left action, since there is only one object in G. Notice that the free orbit F^d is a diagram of sets, or discrete spaces if I is a (non-enriched) small category, but if I is a simplicial category then F^d is a diagram of simplicial sets.

(iii) A *free cellular* I-diagram is a diagram of cofibrant spaces built from free I-cells: $\Delta[n] \times F^d$ by gluing them together along their boundaries $\dot{\Delta}[n] \times F^d$.

Example: If we return to the simplicial map $f : E \to B$ in the above proposition it is not hard to see that $\underset{\sim}{E}$ is a free ΓB-diagram. This follows from a basic property of simplicial complexes: Each point in B belongs to a unique simplex as an interior point. Hence we can construct $\underset{\sim}{E}$ as a diagram of simplicial complexes by induction over the skeletons of B. An illuminating example of the above is the diagram associated with the identity map $f = \mathrm{id} : B \xrightarrow{=} B$ of a simplicial complex. This diagram $\underset{\sim}{B}$: $\Gamma B \to S$ assigns to each simplex $\sigma \in B$ the underlying space of that simplex, and to each face inclusion $\sigma \subset \sigma'$ the inclusion of spaces $\sigma \subset \sigma'$. We get a free diagram where every point $b \in \underset{\sim}{B}(\sigma)$ belongs to a unique free orbit: the orbit F^δ where δ is the unique simplex $\delta \subset \sigma$ such that $b \in \operatorname{int} \delta$.

Free diagram and homotopy colimits

The basic properties of free diagrams are:

PROPOSITION [DF-1]: *For every diagram $\underset{\sim}{X}: I \to S$ there exist a free diagram $\underset{\sim}{X}^{free}$ and natural map $\underset{\sim}{X}^{free} \to \underset{\sim}{X}$ which is a weak equivalence, i.e. such that for all $i \in I$ the induced map $\underset{\sim}{X}_i^{free} \to \underset{\sim}{X}_i$ is a weak equivalence of spaces.*

Proof: (Compare [HH], [D-K-2]) The construction is analogous to that of the CW-approximation to a space and it is done again by induction on the dimension of the free cells $F^d \times \Delta[n]$ in X^{free}. Notice that in this way one gets always a diagram of cofibrant spaces.

PROPOSITION: *For any free diagram $\underset{\sim}{X}^{free}$ the natural map:*

$$\operatorname*{hocolim}_I \underset{\sim}{X}^{free} \xrightarrow{\simeq} \operatorname*{colim}_I \underset{\sim}{X}^{free},$$

from the homotopy colimit to the colimit itself is a weak equivalence.

FREE DIAGRAMS IN THE POINTED CATEGORY. In the pointed category a free-pointed diagram is not free, but it is free away from the base point orbit. Namely it is a diagram constructed from the point diagram over I denoted by $(\underset{\sim}{\star})$ or $(\star_I)$, with $\underset{\sim}{\star}_i = \star$, by adding free cells as above. So it is free relative to $\star_I$.

Recognizing homotopy colimits

Free resolution as above is often an effective method to compute hocolim $\underset{\sim}{X}$.

Example: For any space X with an action of a group G, the map $FG \times X \to X$ is a free G-resolution since $EG \simeq \star$ and the action of G on $EG \times X$ is free. Therefore we get the Borel construction

$$\operatorname{hocolim}_G X \simeq \operatorname*{colim}_G EG \times X = EG \times_G X.$$

Example: Let $\star_I$ be a point diagram. Since as a pointed diagram it is free, the pointed homotopy colimit $\operatorname*{hocolim}_I (\star_I)$ is equivalent to the colimit itself, which is just a point $(*)$. On the other hand, the unpointed homotopy colimit hocolim $\star_I$ is obtained by first constructing a free I-diagram of constructible spaces, denoted here by EI, and since the only map $EI \to *$ is an equivalence on each space in the diagram, one now takes the colimit of EI. In fact one gets: $\operatorname{colim}_I EI = BI$, namely we get the classifying space of I.

Example: If B is any simplicial complex, then $\underset{\sim}{B}$ is a free ΓB-diagram. Therefore

$$\operatorname*{hocolim}_I \underset{\sim}{B} = \operatorname*{colim}_I \underset{\sim}{B} = B$$

since B is clearly obtained by gluing its own simplices along all the simplicial face inclusions.

From diagram to maps

Now for an arbitrary simplicial map $p\colon E \to B$ we have formed a diagram $\underset{\sim}{E}$. From this diagram we can recover not only the space E, but the map p itself up to homotopy:

PROPOSITION: *Any simplicial map $p\colon E \to B$ is equivalent up to homotopy to the map hocolim $\underset{\sim}{E} \to \overline{W}(\Gamma B)$, where $\overline{W}$ denotes the classifying space of a diagram.*

Proof: Since both $\underset{\sim}{E}$ and $\underset{\sim}{B}$ are free ΓB-diagrams we have the equivalences:

$$\begin{array}{ccccc}
\operatorname*{hocolim}_I \underset{\sim}{E} & \xrightarrow{\cong} & \operatorname{colim} \underset{\sim}{E} & \xrightarrow{=} & E \\
\downarrow & & \downarrow & & \downarrow \\
\operatorname*{hocolim}_I \underset{\sim}{B} & \xrightarrow{\cong} & \operatorname{colim} \underset{\sim}{B} & \xrightarrow{=} & B
\end{array}$$

The diagram $\underset{\sim}{E}$ comes with a natural map to the point diagram $*_{\Gamma B}$, so this map induces a map on homotopy colimits. Now the proof is concluded by using the proposition below for $X = B$ and the unfortunate notation $\mathbf{B}$ is the classifying space functor $\overline{W}$:

PROPOSITION: *For any simplicial complex X there is an equivalence*

$$\mathbf{B}\Gamma X \simeq X.$$

Proof: By definition $\mathbf{B}\Gamma X = \operatorname{hocolim}_{\Gamma X}(*)$, the homotopy colimit of the point diagram. But $\underset{\sim}{X} \to *$ is a free resolution of $*$, because it is free, as we saw above, and $\underset{\sim}{X}(\sigma) = \sigma \simeq *$ is a contractible space, being a simplex, so that $\underset{\sim}{X}(\sigma) \to *$ is a weak equivalence for each $\sigma \in \Gamma X$.

Therefore $\operatorname{hocolim}_{\Gamma X} \underset{\sim}{X} \simeq \operatorname{hocolim}_{\Gamma X}(*) \simeq \mathbf{B}\Gamma$, where $\mathbf{B}$ is now the classifying space functor. But we saw that $\operatorname{hocolim}_{\Gamma X} \underset{\sim}{X} \simeq X$, so $X \simeq \mathbf{B}\Gamma X$, as needed.

Homotopy colimits and fibrations, Quillen's Theorem B

We saw above how to write an arbitrary simplicial map $f : E \to X$ in the form $\text{hocolim}_{\Gamma X} \underset{\sim}{E} \to \text{hocolim}_{\Gamma X} *$. Now we specialize to fibre maps. These are distinguished from other maps by the property that the associated diagram $\underset{\sim}{E}$ over the small category ΓX is a diagram of spaces in which all the maps are weak equivalences.

PROPOSITION: *If $E \to X$ is a simplicial fibre map, then for every map $\tau \to \sigma$ in the small category ΓX the map $\underset{\sim}{E}(\tau) \to \underset{\sim}{E}(\sigma)$ is an equivalence.*

Proof: This is the homotopy lifting property that implies the equivalence of fibres along paths in the base space.

Diagrams of fibres computing homotopy fibre

For a simplex $\sigma \in B$ in the base and a simplicial fibre map $E \to X$, the inclusion of the barycenter of σ to σ, namely $b(\sigma) \to \sigma$, induces an equivalence $p^{-1}(b(\sigma)) \hookrightarrow p^{-1}(\sigma) = \underset{\sim}{E}(\sigma)$. Therefore $E(\sigma)$ is equivalent to the fibre over $b(\sigma)$. If X is connected, all these fibres are weakly equivalent.

Conversely, given a diagram of weak equivalences, if one forms the associated map by taking homotopy colimits, one can identify the homotopy fibre of the resulting map as the (constant) homotopy type that appears in the given diagram:

PROPOSITION: *Suppose that I is a small category and let $\underset{\sim}{W}: I \to S$ be a diagram of equivalences, so that for each $i \to j$ in* mor *I the map $\underset{\sim}{W}_i \xrightarrow{\sim} \underset{\sim}{W}_j$ is an equivalence. Assume that the space F is equivalent to all the spaces $\underset{\sim}{W}_i$ for $i \in I$. Then the homotopy fibre of the natural map $\underset{I}{\text{hocolim}}\, \underset{\sim}{W} \to \mathbf{B}I$ is again equivalent to F.*

Proof: This follows by induction on the skeleton from V. Puppe's Theorem above, by decomposing the base. In fact it is also a formulation of Quillen's Theorem B [Q-2].

Relations to homotopy (inverse) limits

Homotopy limits are 'homotopy approximations' to (inverse) limits of a diagram in an analogous way to the relation between colimit and homotopy colimit. Again, these can be defined in various categories. In general, in order to define holim as a functor one chooses once and for all a specific, free resolution of the one-point diagram $(*) = (*_I)$, where $*_I(i) = *$. Let us denote this resolution by $EI \to (*)$ so that EI is a free I-diagram with $EI_i \simeq (pt)$, a contractible space for each $i \in I$. One then takes $\text{holim}_I \underset{\sim}{X}$ for any diagram to be the space of all maps of diagrams

$$\text{holim}_I \underset{\sim}{X} = \underset{I}{\text{Hom}}(EI, \underset{\sim}{X}).$$

It has two crucial properties:

PROPOSITION: *If $f\colon \underset{\sim}{X} \to \underset{\sim}{Y}$ is a weak equivalence of an I-diagrams (i.e. f_i is a weak equivalence for all i) with $\underset{\sim}{X}_i$ and $\underset{\sim}{Y}_i$ fibrant, then the induced map* $\operatorname{holim}_I \underset{\sim}{X} \to \operatorname{holim}_I \underset{\sim}{Y}$ *is a weak equivalence.*

PROPOSITION: *For any diagram $\underset{\sim}{X}$ and any space W, there is a natural weak equivalence*

$$\operatorname{map}(\operatorname{hocolim}_I \underset{\sim}{X}, W) \to \operatorname{holim}_{I^{op}} \operatorname{map}(\underset{\sim}{X}, W).$$

Remark: A special case of the second property is the well-known statement, that if $A \to X \to X/A$ is a cofibration, then for any connected space W the sequence [S]:

$$\operatorname{map}(X/A, W) \to \operatorname{map}(X, W) \to \operatorname{map}(A, W)$$

is a fibre sequence, where $\operatorname{map}(X/A, W)$ is weakly equivalent to the homotopy fibre over the component of the null maps $A \to W$ in $\operatorname{map}(A, W)$. The same statement holds for pointed diagrams with homotopy colimit replaced by pointed homotopy colimit.

Pushout and pullback squares

These diagrams are especially useful examples of both homotopy colimits and homotopy limits. They also have special useful properties that we make explicit now. Given any strictly commutative square:

(Q)
$$\begin{array}{ccc} X & \xrightarrow{f} & Y \\ {\scriptstyle v}\downarrow & & \downarrow{\scriptstyle w} \\ V & \xrightarrow{g} & W \end{array}$$

one can form the homotopy colimit of $V \xleftarrow{v} X \xrightarrow{f} Y$, called also the homotopy pushout, simply by taking the double mapping cylinder $C(f,v)$. This space comes with obvious maps $Y \longrightarrow C(f,v)$ and $V \longrightarrow C(f,v)$. Upon precomposing into X we get another square that is, however, commutative only up to homotopy. From the above square one can also form the strict pushout or identification space as a quotient of Y disjoint union with V. Denote this pushout by PO. As always there is a natural map from the homotopy colimit to the strict colimit. Now by the basic property of the strict direct limit the given square (Q) above induces a natural map $PO \xrightarrow{i} W$. Therefore upon composition one gets a natural map $C(f,v) \longrightarrow W$.

Similar maps exist for the homotopy pullback of the maps $V \longrightarrow W \longleftarrow Y$. This is the usual space of paths $\Lambda(g, w)$. If we denote the strict pullback of the same pair of maps (g, w) by PB. Again there is a natural map $PB \longrightarrow \Lambda(g, w)$. Therefore we get a natural map $X \longrightarrow \Lambda(g, w)$.

Appendix HC: Pointed homotopy coends

Here we discuss briefly a certain kind of homotopy colimit called a homotopy coend; compare [DF-1], [Dw-2], [H-V]. It can be used also to define the homotopy colimit itself.

Given a small category I and two diagrams: an unpointed one over I denoted by U and a pointed one over I^{op} denoted by P, one can form a homotopy version of their coend [Mac]. This version will depend only on the weak homotopy type of these diagrams and is equivalent to the coend itself if one of these diagrams is free as in Appendix HL above. We use the double integral notation to denote this type of pointed homotopy colimit (compare Appendix HL above); this notation is meant to remind the reader that the 'integrand' has two 'variables': One covariant the other contravariant.

DEFINITION: *Let $U : I \to S$ be a functor to (unpointed) spaces and $P : I \to S_*$ be a contravariant functor to pointed spaces. The pointed homotopy coend, denoted*

$$\underset{I}{\mathrm{hocoend}_*}(U, P) = \iint_I U^+ \wedge P = \iint_I U \ltimes P,$$

is defined to be the realization (or diagonal) of the simplicial space $\mathbf{B}(U, I, P)$ which, in dimension k, consists of the one point wedge:

$$\bigvee_{c_0 \to c_1 \to \cdots \to c_k} U^+(c_0) \wedge P(c_k),$$

indexed by the k-simplices $c_0 \to c_1 \to \cdots \to c_k$ of the nerve of I. The face and degeneracy maps in the simplicial space $\mathbf{B}(U,I,P)$ are the usual ones as follows: Let $f_n \circ \cdots \circ f_1 \colon \alpha \to \beta$ be composeable, and let $p \in P_\alpha, u \in U_\beta$, one has:

$$d^i(u; f_1, f_2 \cdots, f_n; p) = \begin{cases} (u; f_1, \cdots, f_{n-1}; f_n(p)), & i = 0 \\ (u; f_1, \cdots, f_i \circ f_{i+1} \cdots, f_n; p)), & 1 \le i \le n-1 \\ (f_1(u); f_2, \cdots, f_n; p)), & i = n. \end{cases}$$

The degeneracies are defined by insertion of identity maps.

WARNING: Our notation regarding I and I^{op} is opposite to the one used by [H-V].

Free resolution approach Compare [H-V, 3.2]. Just as in the case of homotopy colimit in appendix HL above, here too one can reduce the notion of homotopy coend to that of (strict) coend $- \otimes_I -$ by first resolving one of the two diagrams involved by free diagram.

Recall (Appendix HL) that for a free diagram the notions of homotopy colimit and (strict) colimit are canonically equivalent. Analogously, in case the diagram P above is free, the strict coend is equivalent to the homotopy coend. To see that one shows by a direct computation that it is so for any free orbit over I^{op} and then one proceeds by induction on the skeleta of P using the fact that both the coend and the homotopy coend commutes with the pushouts used to attach free cells $\Delta[n] \times F^i$. To conclude, let us denote as usual the (strict) coend by the tensor product notation $\otimes_I$; we have:

Proposition: *If $U : I \longrightarrow S$ (Res. $P : I^{op} \longrightarrow S_*$) is free (res. pointed free) then the natural map*

$$\iint_I U^+ \wedge P \longrightarrow U^+ \otimes_I P$$

is a weak equivalence. ∎

Thus we have

$$\iint_I U^+ \wedge P \simeq U^+ \otimes_I P^{free}.$$

In this way we see that the homotopy coend is the total derived functor of coend or tensor product of two diagrams.

A generalization: We also use a generalization of this concept when, instead of the smash product between two pointed spaces, one uses any binary homotopy functor, say $W\#V$, such as the join $W * V$. In that case we denote by

$$\underset{I}{\text{hocoend}_*}(U, P; \#) = \iint_I U\#P$$

the diagonal of the corresponding simplicial space $\mathbf{B}(U, I, P; \#)$, which in dimension k consists of the one point wedge

$$\bigvee_{c_0 \to c_1 \to \cdots \to c_k} U(c_0)\#P(c_k)$$

indexed by the k-simplices $c_0 \to c_1 \to \cdots \to c_k$ of the nerve of I. The face and degeneracy maps in the simplicial space $\mathbf{B}(U, I, P; \#)$ are the usual ones.

It follows from a direct inspection of the construction of a pointed homotopy colimit in [B-K] that one can write the pointed homotopy colimit as a coend:

PROPOSITION: *There is a natural equivalence for any diagram of pointed spaces P as above:*

$$\operatorname*{hocolim}_{I}{}_* P \simeq \operatorname*{hocoend}_{I}{}_*(*, P).$$

EXAMPLE: G-SPACE AS A (HOMOTOPY) COEND Given a <u>pointed</u> G-space $* \in X$, one can recover it from the diagram of the fixed points as a coend as explained in [DF-1] and [D-K-2]. In this case the coend is equivalent to the homotopy coend since we get a free diagram as an input (see below). Namely we associate with X two diagrams over the categories O, O^{op} of G-orbits: O is the category of G-orbits, i.e. the G spaces G/K, for K any subgroup of G where the maps are G-equivariant maps. First, the identity functor $\mathcal{I} = \mathcal{I}_G$ that assigns to each G-orbit of the form G/H, that orbit itself as a G-space. This is a functor $\mathcal{I}\colon O \to \mathcal{S}$ from O to unpointed spaces, since none of these orbits except the trivial one G/G has a G-fixed point. Second, the fixed-point assignment is a contravariant functor denoted by $X^O\colon O \to \mathcal{S}_*$ that assigns the space $X^K = \mathrm{map}_G(G/K, X)$ to the orbit G/K and maps coming from composition on the domain in the mapping space presentation of the fixed-point set. Notice that now the coend itself has a natural structure of a G-space coming from the left action of G on the orbits that enter into the coend.

There is an evaluation map $\mathrm{coend}(\mathcal{I}, \mathrm{X}^{\mathrm{O}}) \longrightarrow \mathrm{X}$ which takes the summand $G/H \times \mathrm{map}_G(G/K, X)$ that corresponds to a specific map $\omega\colon G/H \to G/K$ via the obvious composition and evaluation. We now quote from standard G-space theory :

PROPOSITION: *The evaluation map gives a G-equivalence:* $\mathrm{coend}(\mathcal{I}, \mathrm{X}^{\mathrm{O}}) \xrightarrow{\simeq} \mathrm{X}$ *form the coend to X.*

Outline of Proof: The main point is to notice that the functor $\mathrm{map}_G(G/L, --)$ on G spaces where L is a subgroup of G commutes with homotopy colimits up to equivalence, because G/L is an 'orbit' so it is 'small'. So one shows that the evaluation map induces an equivalence on L-fixed-point sets for any $L \subseteq G$. ∎

References

[A] J.F. Adams, Infinite Loop Spaces, Princeton University Press.

[Am] A. Amit, *Direct limits over categories with contractible nerve*, Master Thesis, The Hebrew University of Jerusalem, 1994.

[An] D. Anderson, *Localizing CW complexes*, Ill. J. Math. **16**(1972) 519–525.

[An-H] D. Anderson and L. Hodgkin, *The K-theory of Eilenberg–Mac Lane complexes*, Topology **7**(1968) 317–329.

[B-1] A.K. Bousfield, *Localization of spaces with respect to homology*, Topology **14**(1975), 133–150.

[B-2] A.K. Bousfield, *Construction of factorization systems in categories*, J. Pure Appl. Alg. **9**(1976) 207–220.

[B-3] A.K. Bousfield, *K-localizations and K-equivalences of infinite loop spaces*, Proc. London Math. Soc. **44**(1982) 291–311.

[B-4] A.K. Bousfield, *Localization and periodicity in unstable homotopy theory*, J. Am. Math. Soc. **7**, No. 4(1994) 831–874.

[B-5] A.K. Bousfield, *Unstable localizations and periodicity*, Proceedings of the Barcelona Conference on Algebraic Topology, 1994 (to appear).

[B-6] A.K. Bousfield, *On homology equivalences and homological localizations of spaces*, Am. J. Math. **104**(1982), 1025–1042.

[Baum] G. Baumslag, *Some aspects of groups with unique roots*, Acta Math. **104** (1960) 217–303.

[Baum] G. Baumslag, *Roots and wreath products*, Proc. Camb. Phil. Soc. **56**(1960) 109–117.

[B-F] A.K. Bousfield and E.M. Friedlander, *Homotopy theory of Γ-spaces, spectra, and bisimplicial sets*, Springer Lecture Notes in Math. **658**(1978) 80–130.

[B-K] A. K. Bousfield and D. M. Kan, *Homotopy Limit, Completions, and Localizations*, Springer Lecture Notes in Math. **304**(1972).

[Bl-1] D. Blanc, *Loop spaces and homotopy operations*, preprint (1995).

[Bl-2] D. Blanc, *Mapping spaces and M-CW complexes*, preprint (1995).

[Ca] C. Casacuberta, *Anderson localization from a modern point of view*, Proceeding AMS Conference 1993, Northeastern University Contemp. Math. **181**(1995) 35–44.

[Ca-P] C. Casacuberta and G. Peshke, *Localizing with respect to self-map of the circle*, Trans. Am. Math. Soc. **339**(1993) 117–140.

[Ca-R] C. Casacuberta and J.L Rodriguez, *On weak homotopy equivalence between certain mapping spaces*, preprint.

[C-E] H. Cartan and S. Eilenberg, Homological Algebra, Princeton University Press, 1956.

[Ch-1] W. Chacholsky, *Homotopy properties of shapes of diagrams*, Report No. 6, Institute Mittag-Leffler, 1993/94.

[Ch-2] W. Chacholsky, *On the functors CW_A and P_A*, Ph.D. Thesis, Notre Dame University, preprint, 1994.

[Ch-3] W. Chacholsky, *Closed Classes*, Proceedings of the Barcelona Conference on Algebraic Topology, 1994 (to appear).

[Ch-4] W. Chacholsky, *A generalization of the triad theorem of Blakers–Massey*, Preprint, (1995).

[C-N] F. Cohen and J. Neisendorfer, *A note on de-suspending the Adams map*, Math. Proc. Camb. Phil. Soc. **99**(1968) 59–64.

[Co] M.M. Cohen, A Course in Simple-Homotopy Theory, GTM series, Springer-Verlag, Berlin, 1973.

[C-P-P] C. Casacuberta, G. Peshke, and M. Pfenniger, *On orthogonal pairs in categories and localization*, London Math. Soc. Lecture Notes Ser. **175**(1992) 211–223.

[Del] A. Deleanu, *Existence of Adams completions for objects in cocomplete categories*, J. Pure Appl. Alg. **6**(1975) 31–39.

[DF-1] E. Dror Farjoun, *Homology and homotopy of diagrams of spaces*, Algebraic Topology, Proceedings, Seattle Conference 1985, Springer Lecture Notes in Math. **1286**(1987) 93–134.

[DF-2] E. Dror Farjoun, *The localization with respect to a map and v_1-periodicity*, Proceedings of the Barcelona Conference 1990, Springer Lecture Notes in Math. **1509**(1992) 104–113.

[DF-3] E. Dror Farjoun, *Localizations, fibrations and conic structures*, preprint.

[DF-4] E. Dror Farjoun, *Higher homotopies of natural construction*, J. Pure Appl. Alg. (to appear).

[DF-5] E. Dror Farjoun, *Cellular inequalities*, Proceeding AMS Conference 1993, Northeastern University, Contemp. Math. **181**(1995) 159–181.

[DF-6] E. Dror Farjoun, *Acyclic spaces*, Topology **11**(1972) 339–348.

[DF-S] E. Dror Farjoun and J. H. Smith, *Homotopy localizations nearly preserves fibrations*, Topology **34** , No. 2(1995) 359–375.

[DF-Z-1] E. Dror Farjoun and A. Zabrodsky, *Unipotency and nilpotency in homotopy equivalences*, Topology **18**(1979) 187–197.

[DF-Z-2] E. Dror Farjoun and A. Zabrodsky, *Homotopy equivalences in diagrams of spaces*, J. Pure Appl. Alg. **45**(1986) 169–182.

[D-H-S] E. Devinatz, M. J. Hopkins, and J. H. Smith, *Nilpotence and stable homotopy*, Ann. of Math. **128**(1988) 207–242.

[D-K-1] W. Dwyer and D.M. Kan, *Simplicial localization*, Topology **19**(1980) 427–440.

[D-K-2] W. Dwyer and D.M. Kan, *Singular functors and realization functors*, Proceed Nederland Acd. Sci. A, **87**(1984) 147–153.

[D-K-3] W. Dwyer and D.M. Kan, *Function complexes for diagrams of simplicial sets*, Indag. Math. **45**(2)(1983) 139–147.

[D-Th] A. Dold and R. Thom, *Quasifibration and infinite symmetric products*, Ann. of Math. (2) **67**(1958) 239–281.

[Dw-1] W. Dwyer, *The centralizer decomposition of BG*, Proceedings of the Barcelona Conference in Algebraic Topology 1994 (to appear).

[Dw-2] W. Dwyer, *Symmetric products*, Draft (1993).

[G] T. Ganea, *A generalization of homology and homotopy suspension*, Comment. Math. Helv. **39**(1965) 295–322.

[H] V. Halperin, *Equivariant localization*, preprint.

[He] A. Heller, *Homotopy theories*, Memoirs Am. Math. Soc. **383**(1988).

[HFLT] S. Halperin, Y. Felix, J.-M. Lemaire and J.C. Thomas, Mod-p loop space homology, Invent. Math. **95** (1989) 247–262.

[HH] P. Hirschhorn, *Localization in model categories* (in preparation).

[Ho] M.J. Hopkins, *Global methods in stable homotopy theory*, London Math. Soc. Lecture Note Ser. **117**(1987) 73–96.

[H-S] M. J. Hopkins and J. H. Smith, *Nilpotence and stable homotopy theory II*, Ann. of Math. (to appear).

[H-V] J. Hollender and R.M. Vogt, *Modules of topological spaces, applications to homotopy limits and E_∞ structures*, Arch. Math. **59**(1992) 115–129.

[J] I.M. James, *Reduced product spaces*, Ann. of Math.(2) **62**(1955) 170–197.

[J-M-O] S. Jackowsky, J.E. MacClure, and R. Oliver, *Homotopy classification of self maps of BG via G-actions*, Ann. of Math. **135**(1992) 227–270.

[L-T-W] D.M. Latch, R.W. Thomason, and W.S. Wilson, *Simplicial sets from categories*, Math. Z. **164**(1979) 195–214.

[M] H. Miller, *The Sullivan conjecture about maps of $K(\pi, 1)$ to finite complexes*, Ann. of Math. (2) **120**(1984) 39–87.

[Mac] S. Mac Lane , *Categories for the working mathematician*, GTM 5, Springer Verlag, Berlin, 1971.

[May-1] J.P. May, *Simplicial objects in algebraic topology*, Van Nostrand Math. Studies **11**(1967).

[May-2] J.P. May, *Fibrewise localization and completion*, Trans. Am. Math. Soc. **258**, No.1 (1980) 127–146.

[McG] C.A. McGibbon, *Infinite loop spaces and Neisendorfer localization*, preprint (1995)

[Mis] G. Mislin, *On localization with respect to K-theory*, J. Pure Appl. Alg. **10**(1975) 201–213.

[Mit] S.A. Mitchell, *Finite complexes with $A(n)$-free cohomology*, Topology **24** (1985) 227–246.

[Mo] J. Moore, *Semi-simplicial complexes and Postnikov systems*, Symp. International Topological Algebra, Univ. Autonoma de Mexico (1958) 232–247.

[M-R] M. Mahowald and W. Richter, *There exists a $v_1{}^4$ Adams map $M^{13} \longrightarrow M^5$ of James filtration two* (to appear).

[M-Rav] M. Mahowald and D. Ravenel, *Toward a global understanding of stable homotopy groups of sphere*, Contemp. Math. **58 II**(1987) 75–88.

[M-S] J. Moore and L. Smith, *Hopf algebra and multiplicative fibrations I*, Am. J. Math. **90**(1968) 752–780.

[M-T] M. Mahowald and R. Thompson, *K-theory and unstable homotopy groups*, Contemp. Math. **96**(1989) 273–279.

[N] J. Neisendorfer, *Localizations and connected covers of finite complexes*, Proceeding AMS meeting, Northeastern Univ. (1993), Contemp. Math. **181**(1995).

[No] A. Nofech, *Localization of homotopy limits*, Ph.D. Thesis, The Hebrew University of Jerusalem, 1992.

[Oka] S. Oka, *Existence of the unstable Adams map*, Mem. Fac. Science Kyushu Univ. **42**(1988) 95–108.

[Pe] G. Peschke, *Localizing groups with action*, Publ. Math. Univ. Autonoma, Barcelona **33**(1989) 227–234.

[Pu] V. Puppe, *A remark on homotopy fibrations*, Manuscr. Math. **12** (1974) 113–120.

[Q-1] D.G. Quillen, Homotopical Algebra, Springer Lecture Notes in Math. **43** (1967).

[Q-2] D.G. Quillen, *Higher algebraic K-theory: I, Algebraic K-theory I*, Springer Lecture Notes in Math. **341**(1973) 85–147.

[R] D.C. Ravenel, Nilpotency and Periodicity in Stable Homotopy Theory, Princeton University Press, New York, 1992.

[R-W] D.C. Ravenel and W.S. Wilson, *The Morava K-theory of Eilenberg–Mac Lane spaces and the Conner-Floyd conjecture*, Am. J. Math. **102**(1980) 691–748.

[S] G. Segal, *Categories and cohomology theories*, Topology **13**(1974) 293–312.

[Sm-1] J. H. Smith, *Finite complexes with vanishing line of small slope*, Israel J. Math. (to appear).

[Sm-2] J. H. Smith, Private communication, 1992.

[Sp] E. Spanier, Algebraic Topology, McGraw-Hill, New York, 1966.

[S-V] R. Schwanzl and R.M. Vogt, *E_∞-monoid with coherent homotopy inverses are abelian groups*, Topology **28**(1989) 481–484.

[T] R. Thompson *A relation between K theory and unstable homotopy groups with application to $B\Sigma_p$*, Contemp. Math. **146**(1993) 431–439.

INDEX

August 1995. The Hebrew University of Jerusalem.
e-mail address: farjoun@sunset.huji.ac.il (emm)

Vol. 1574: V. P. Havin, N. K. Nikolski (Eds.), Linear and Complex Analysis – Problem Book 3 – Part II. XXII, 507 pages. 1994.

Vol. 1575: M. Mitrea, Clifford Wavelets, Singular Integrals, and Hardy Spaces. XI, 116 pages. 1994.

Vol. 1576: K. Kitahara, Spaces of Approximating Functions with Haar-Like Conditions. X, 110 pages. 1994.

Vol. 1577: N. Obata, White Noise Calculus and Fock Space. X, 183 pages. 1994.

Vol. 1578: J. Bernstein, V. Lunts, Equivariant Sheaves and Functors. V, 139 pages. 1994.

Vol. 1579: N. Kazamaki, Continuous Exponential Martingales and *BMO*. VII, 91 pages. 1994.

Vol. 1580: M. Milman, Extrapolation and Optimal Decompositions with Applications to Analysis. XI, 161 pages. 1994.

Vol. 1581: D. Bakry, R. D. Gill, S. A. Molchanov, Lectures on Probability Theory. Editor: P. Bernard. VIII, 420 pages. 1994.

Vol. 1582: W. Balser, From Divergent Power Series to Analytic Functions. X, 108 pages. 1994.

Vol. 1583: J. Azéma, P. A. Meyer, M. Yor (Eds.), Séminaire de Probabilités XXVIII. VI, 334 pages. 1994.

Vol. 1584: M. Brokate, N. Kenmochi, I. Müller, J. F. Rodriguez, C. Verdi, Phase Transitions and Hysteresis. Montecatini Terme, 1993. Editor: A. Visintin. VII. 291 pages. 1994.

Vol. 1585: G. Frey (Ed.), On Artin's Conjecture for Odd 2-dimensional Representations. VIII, 148 pages. 1994.

Vol. 1586: R. Nillsen, Difference Spaces and Invariant Linear Forms. XII, 186 pages. 1994.

Vol. 1587: N. Xi, Representations of Affine Hecke Algebras. VIII, 137 pages. 1994.

Vol. 1588: C. Scheiderer, Real and Étale Cohomology. XXIV, 273 pages. 1994.

Vol. 1589: J. Bellissard, M. Degli Esposti, G. Forni, S. Graffi, S. Isola, J. N. Mather, Transition to Chaos in Classical and Quantum Mechanics. Montecatini Terme, 1991. Editor: S. Graffi. VII, 192 pages. 1994.

Vol. 1590: P. M. Soardi, Potential Theory on Infinite Networks. VIII, 187 pages. 1994.

Vol. 1591: M. Abate, G. Patrizio, Finsler Metrics – A Global Approach. IX, 180 pages. 1994.

Vol. 1592: K. W. Breitung, Asymptotic Approximations for Probability Integrals. IX, 146 pages. 1994.

Vol. 1593: J. Jorgenson & S. Lang, D. Goldfeld, Explicit Formulas for Regularized Products and Series. VIII, 154 pages. 1994.

Vol. 1594: M. Green, J. Murre, C. Voisin, Algebraic Cycles and Hodge Theory. Torino, 1993. Editors: A. Albano, F. Bardelli. VII, 275 pages. 1994.

Vol. 1595: R.D.M. Accola, Topics in the Theory of Riemann Surfaces. IX, 105 pages. 1994.

Vol. 1596: L. Heindorf, L. B. Shapiro, Nearly Projective Boolean Algebras. X, 202 pages. 1994.

Vol. 1597: B. Herzog, Kodaira-Spencer Maps in Local Algebra. XVII, 176 pages. 1994.

Vol. 1598: J. Berndt, F. Tricerri, L. Vanhecke, Generalized Heisenberg Groups and Damek-Ricci Harmonic Spaces. VIII, 125 pages. 1995.

Vol. 1599: K. Johannson, Topology and Combinatorics of 3-Manifolds. XVIII, 446 pages. 1995.

Vol. 1600: W. Narkiewicz, Polynomial Mappings. VII, 130 pages. 1995.

Vol. 1601: A. Pott, Finite Geometry and Character Theory. VII, 181 pages. 1995.

Vol. 1602: J. Winkelmann, The Classification of Three-dimensional Homogeneous Complex Manifolds. XI, 230 pages. 1995.

Vol. 1603: V. Ene, Real Functions – Current Topics. XIII, 310 pages. 1995.

Vol. 1604: A. Huber, Mixed Motives and their Realization in Derived Categories. XV, 207 pages. 1995.

Vol. 1605: L. B. Wahlbin, Superconvergence in Galerkin Finite Element Methods. XI, 166 pages. 1995.

Vol. 1606: P.-D. Liu, M. Qian, Smooth Ergodic Theory of Random Dynamical Systems. XI, 221 pages. 1995.

Vol. 1607: G. Schwarz, Hodge Decomposition – A Method for Solving Boundary Value Problems. VII, 155 pages. 1995.

Vol. 1608: P. Biane, R. Durrett, Lectures on Probability Theory. VII, 210 pages. 1995.

Vol. 1609: L. Arnold, C. Jones, K. Mischaikow, G. Raugel, Dynamical Systems. Montecatini Terme, 1994. Editor: R. Johnson. VIII, 329 pages. 1995.

Vol. 1610: A. S. Üstünel, An Introduction to Analysis on Wiener Space. X, 95 pages. 1995.

Vol. 1611: N. Knarr, Translation Planes. VI, 112 pages. 1995.

Vol. 1612: W. Kühnel, Tight Polyhedral Submanifolds and Tight Triangulations. VII, 122 pages. 1995.

Vol. 1613: J. Azéma, M. Emery, P. A. Meyer, M. Yor (Eds.), Séminaire de Probabilités XXIX. VI, 326 pages. 1995.

Vol. 1614: A. Koshelev, Regularity Problem for Quasilinear Elliptic and Parabolic Systems. XXI, 255 pages. 1995.

Vol. 1615: D. B. Massey, Lê Cycles and Hypersurface Singularities. XI, 131 pages. 1995.

Vol. 1616: I. Moerdijk, Classifying Spaces and Classifying Topoi. VII, 94 pages. 1995.

Vol. 1617: V. Yurinsky, Sums and Gaussian Vectors. XI, 305 pages. 1995.

Vol. 1618: G. Pisier, Similarity Problems and Completely Bounded Maps. VII, 156 pages. 1996.

Vol. 1619: E. Landvogt, A Compactification of the Bruhat-Tits Building. VII, 152 pages. 1996.

Vol. 1620: R. Donagi, B. Dubrovin, E. Frenkel, E. Previato, Integrable Systems and Quantum Groups. VIII, 488 pages. 1996.

Vol. 1621: H. Bass, M. V. Otero-Espinar, D. N. Rockmore, C. P. L. Tresser, Cyclic Renormalization and Automorphism Groups of Rooted Trees. XXI, 136 pages. 1996.

Vol. 1622: E. D. Farjoun, Cellular Spaces, Null Spaces and Homotopy Localization. XIV, 199 pages. 1996.